Contents

Introduction *iv*

Acknowledgements *iv*

Hints on studying and revision *v*

Practical skills *vi*

Exams *viii*

Section 1   INTERACTIONS
1. Forces 2
2. Measuring forces 13
3. Forces in balance 17
4. Measuring motion 26
5. Force and motion 38
   Forces and Motion check list 49
   Revision quiz 50
   Examination questions 51
6. Pressure 53
   Pressure check list 67
   Revision quiz 68
   Examination questions 69

Section 2   ENERGY
7. Energy 72
8. Energy transfer 82
9. Power 86
10. Machinery 90
11. Wave energy 95
12. Waves we can hear 104
    Energy and Waves check list 113
    Revision quiz 114
    Examination questions 116
13. Light energy 119
14. Light rays 123
    Light and Images check list 138
    Revision quiz 139
    Examination questions 140
15. Heat at work 141
16. The expansion of solids 148
17. The conduction of heat 159

18. The radiation of heat 162
19. Temperature 169
    Heat energy check list 172
    Revision quiz 173
    Examination questions 175
20. Electric energy 178
    Electric charge check list 193
    Revision quiz 194
21. Cells and voltage 195
22. Electric resistance 207
23. Electric power 227
24. Electricity in the home 233
    Electric energy in circuits check list 241
    Revision quiz 242
    Examination questions 243
25. Magnetism 246
26. Electromagnetism 252
27. The electric motor force 260
28. Electricity generators 263
29. Electric transformers 273
    Electromagnetism check list 281
    Revision quiz 282
    Examination questions 283
30. Electronics—the transistor 285
31. Electronics—microchips that make decisions 289
    Electronics check list 293
    Revision quiz 294
    Examination questions 295

Section 3   MATTER
32. Measuring matter 298
33. The change of state 303
34. The kinetic theory of matter 314
35. Radioactivity and the atom 320
    Matter check list 334
    Revision quiz 336
    Examination questions 338

Answers 340

Index 342

Introduction

This book is designed for use throughout a two- or three-year course leading to a GCSE examination in Physics. It also covers the physics required by most Combined and Integrated science courses at this level.

A commonsense approach is used that focuses on familiar events and develops a detailed understanding of what is going on. Ideas are taken to two levels that can be followed according to ability and interest. The levels (core and extension) are clearly distinguished, making the book suitable for building a foundation in Physics and sufficient for the future specialist. 'Core' level work has been given red headings and lies beneath red rulings. 'Extension' material is separated by black rulings and has black headings.

The material is designed to be both active and attractive to use. Each new idea is introduced carefully by words and pictures and investigated by simple experiments. Possible outcomes of observations are included for those who are not able to do, or observe, the experimental work. Active involvement with equipment and learning from first-hand experience are implicit in the presentation and the development of practical skills is an important outcome of studying the subject. Hints on identifying these skills are included and experiments suitable for assessing practical skills have been identified by an ●.

Large numbers of structured questions are built into the text to test and extend understanding. At intervals there are check lists and matching quizzes that identify a body of knowledge for examination purposes. The check lists together effectively outline the examination syllabus. The check lists are followed by sample questions of the style and level of GCSE examination questions.

The order of presentation represents one logical route through the material. However the breakdown of topics will permit a wide variety of teaching routes and uses. Examination and revision hints are included to help students organize and consolidate essential information.

To approach Physics by focusing common sense on the familiar leads naturally to the inclusion of many applications of physics. These inventions effect the quality of our lives and can be used to promote debate on the social implications of scientific discoveries.

Many find language a barrier to understanding and the level of language used here has been most carefully monitored. Diagrams, photographs and cartoons complement the words, conveying their meaning in a more immediate way. The pictures also show Physics as a human – and so humorous – involvement with the (sometimes) puzzling and unexpected. The book aims to help its users enjoy grappling with scientific ideas, so that the Physics they learn will make a useful contribution to their life experience.

I am very grateful to Dave Williams for his ideas and advice on technical matters and to the many pupils who inadvertently helped me in the preparation of this book.

Peter Warren

Acknowledgements

The publisher is grateful to the following examination boards for giving permission to reproduce questions: Northern Examining Association (comprising Associated Lancashire Schools Examining Board, Joint Matriculation Board, North Regional Examinations Board, North West Regional Examinations Board, Yorkshire and Humberside Regional Examinations Board), Southern Examining Group, Midland Examining Group, and the London and East Anglian Group.

Photographs

Colorsport cover, p. 86; Barnaby's Picture Library p. 1, 102, 126, 156, 297, 322 (Tony Boxall); British Rail p. 28; Oxford Scientific Films Ltd p. 71; David Redfern Photography p. 74; Water Authorities Association p. 78 (left); J. Allan Cash Photo Library p. 78 (top right), 79; Daily Telegraph Colour Library p. 78 (centre right); Ardea p. 78 (bottom right); from *Ripple Tank Studies* by W. Llowarch © Oxford University Press 1961 p. 99; British Telecommunications p. 101; David Williams p. 121, 195, 257, 261, 285; Charing Cross and Westminster Medical School p. 132; British Airways p. 142; Howard Jay p. 157; Fire Research Station p. 168; Central Electricity Generating Board p. 279; Neville Fox-Davies/Bruce Coleman p. 306; (from *PSSC Physics*) reproduced by permission of D C Heath and Company, © 1965 p. 314; Sygma/John Hillelson Agency p. 329; Science Photo Library p. 330; Associated Press p. 333.

Hints on studying and revision

It is important to learn to study. Here are a few ideas that may help you.

Where?
Find a quiet place on your own if you can. Get a table for your books and a good light. Studying should be active, so have a pencil and paper ready for making notes and testing yourself.

When?
Study when you are not too tired, and for short periods at a time. Plan to work in short bursts (say half-an-hour sessions) with a break in between.

How?
You should organize your work in the way that suits you best, but here is one way you can study with this book.

Make a revision notebook (or revision cards)

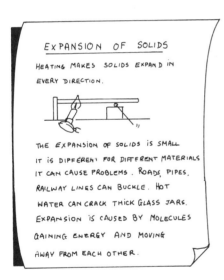

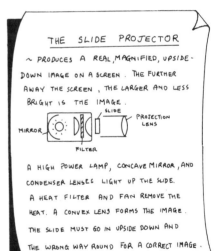

1. Turn to a page and write down the topic title.

2. Draw a diagram of apparatus used to study that topic.

3. Read the page and write a list of facts about the topic.

Most of the topics can be summarized in this way, allowing you to build up a revision notebook and to learn at the same time. You should then go through these notes a number of times until you can remember them.

Practical skills

You learn most about Physics by doing experiments. They bring you face to face with the physical world so that you can see for yourself how it behaves.

Doing an experiment requires practical skills. You have to be able to choose and assemble apparatus, make observations (or measurements) and draw conclusions from your results. These skills develop as you experiment and are an important part of the study of Physics.

The practical skills needed to do an experiment can be grouped into four main types. They are listed here so that you can check your progress as you study.

Skill A. Handling apparatus

Can you choose suitable apparatus for an experiment?
Can you put apparatus together properly so that it works safely?
Can you follow instructions on how to use a piece of apparatus?
Can you follow instructions on how to do an experiment?

Examples of skill A
- setting up a retort stand
- pouring a liquid between containers
- operating a lamp from a power supply
- connecting an ammeter and voltmeter
- using goggles when heating things
- following instructions for making a speed–time graph from a ticker-tape
- putting apparatus together from a picture
- making a circuit from a circuit diagram
- using a bunsen burner correctly

Skill B. Measuring and Observing

Can you read the scales of instruments?
Can you check and allow for the zero error of an instrument?
Do you know when to make repeat measurements?
Can you make rough estimates of quantities?
Are you able to make careful observations of events?

Skill C. Recording

Can you record results in neat tables?
Can you record results in a sensible order?
Can you record the units of results?

Examples of skill B
- measuring spring extension with a metre rule
- measuring current and voltage scales
- reading a thermometer
- measuring volume with a measuring cylinder
- measuring mass on a balance
- measuring time intervals with a clock
- observing events when water boils
- observing how lights are controlled by switches or transistors
- observing forces between magnets
- observing voltage waves on an oscilloscope

Can you recognize mistaken readings?
Can you describe your observations clearly?
Can you plot results on a graph using suitable scales?
Can you label and head your graphs correctly?
Can you take readings from your graphs?
Can you use a graph to show that two quantities are proportional?

Skill D. Designing an experiment

Can you design an experiment to measure a quantity?
Can you choose the apparatus you need?
Can you plan how it should be put together?
Can you design an experiment to find how one quantity depends on another?
Can you plan an experiment to find the answer to a question?

All experiments in Physics require the use of some or all of these skills. Some experiments, however, are particularly useful for testing the four types of skill and have been marked with ◐.

Examples of skill D

designing an experiment to
- measure the resistance of an iron wire
- measure the work done by a motor in lifting a weight
- measure the average speed of a passing car
- measure the electrical power of a motor
- measure the melting and freezing point of wax
- find out how the resistance of a thermistor changes with temperature
- find out how the voltage of a dynamo changes with its speed
- find out how the secondary voltage of a transformer depends on its number of turns
- find out what type of surface radiates most heat
- find out which liquid would be best for cooling an engine

Exams

If you are properly prepared you can even enjoy taking exams. Proper preparation will also ensure that you do your best and will show the examiners how good you really are. Here are some points to bear in mind as exams draw near.

Well before

1. Start revision some months before the exams. Make brief notes from your notebooks and textbooks (see p. *v*). Go over the work until you can remember it.

2. Try questions from past papers. Practise all the different types of question that are asked. Work out how many minutes you have to answer each question and write out answers in that time. Learn to answer exactly what is asked. Putting ideas into words is difficult so take every opportunity to practise writing. Use drawings to help whenever you can.

Just before

Double-check the dates and times of examinations. Arrive fresh and in good time to avoid last minute panic. Have two pens, a pencil, rubber, a watch, a calculator, a ruler and drawing instruments if they are needed.

During

Read the instructions carefully and ask if you do not understand them. If there is a choice, choose quickly but carefully, doing your best question first. Keep an eye on the clock so there is time to answer the right number of questions. At the end check your arithmetic, spelling and look for careless slips.

Good luck!

After...

Section 1
INTERACTIONS

1 Forces

We are always pushing, pulling and lifting. A push or a pull is called a **force**.

A force can get an object moving, or stop it from moving, or change the direction of its movement, or squeeze it and change its shape. Force is measured in **newtons** (N). A force of one newton is quite a small force. The drawings show people exerting forces and give a rough idea of the size of those forces.

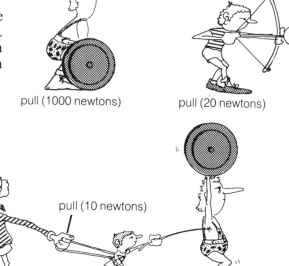

1 Here are some forces in action. For each picture:
   (a) choose one of the forces that is acting,
   (b) name the object that the force acts on,
   (c) say what the force is doing by choosing from the following list: getting an object moving; stopping an object that is moving; changing the direction of motion; balancing another force and preventing movement; stretching an object; bending an object.

2 In sport, music and dancing we learn to use forces with great skill. We enjoy using force to get exactly the result we want. These pictures show forces being used with precision and skill. The sizes of the forces are also shown in the pictures.

Make a table with the following headings and complete it for all the pictures:

Size of the force	What applies the force	What the force does

List the forces in order of size.

Force vectors

There is no body on Earth that is free from all force and although we are used to feeling and using forces, they are not easy to visualize. One way of making forces more obvious is to draw them as force **vectors**.

A force vector is a line with an arrow at one end that **starts** at the place **where the force acts**. The **length** of the line shows the **size** of the force and the **arrow** points in the **direction** of the force.

The pattern of forces that act in a situation can be complex. It helps to extract a small part of the action and draw in all the force vectors on that part. This is then known as a **free-body diagram**. The pictures in the circles are examples of free-body diagrams.

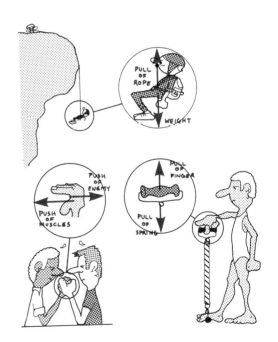

3 Draw free-body diagrams showing all the forces that act on:
   (a) a spider hanging on a thread,
   (b) a finger pulling on a bicycle brake,
   (c) a cat's tail being trodden on,
   (d) an apple resting on a table.
   Give a name to each of the forces you have marked.

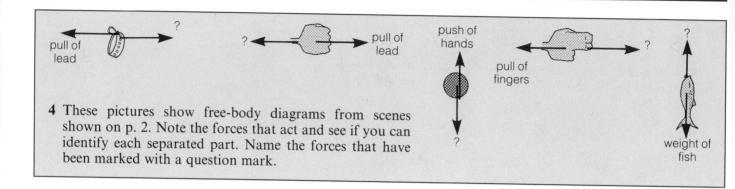

4 These pictures show free-body diagrams from scenes shown on p. 2. Note the forces that act and see if you can identify each separated part. Name the forces that have been marked with a question mark.

Reaction forces

You will notice that when a force presses on a surface, that surface pushes back. The force of the surface is sometimes called the **reaction** to the force that caused it – the **action**. These examples show how an action force always produces a reaction.

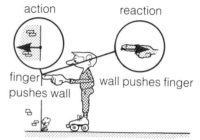

The skater pushes on the wall and the reaction force from the wall pushes back on his finger. The reaction force accelerates him away from the wall.

The rower pushes against the firm bank and the bank pushes back on his pole. The reaction force moves the boat across the water.

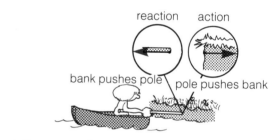

The mother pulls the stubborn child and the unexpected size of the reaction force tips her off balance.

In all such cases the action and reaction forces act on **different** bodies. They are also equal in size and act in opposite directions.

Newton's third law of motion

This law summarizes the way that action and reaction forces behave. It says that:

'Action and reaction are equal, opposite and act on different bodies.'

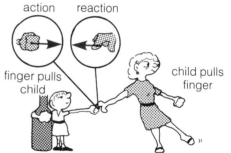

5 What would happen if action and reaction forces were not equal?

6 Two lads on roller skates hold on to either end of a rope. One lad pulls the rope through his hands; the other holds on.
  (a) Describe how they move.
  (b) Draw free-body diagrams for: (i) the hand that pulls, (ii) the hand that holds on, (iii) the rope.
  (c) Label the forces 'action' or 'reaction'.

The force of gravity

There is a small force between any two bodies that pulls them together. Usually this force is very small, but if one of the bodies is the Earth, the force (close to) is far from small. Wherever we are on the Earth, a strong force pulls us to its surface. We call this the force of gravity.

The huge mass of the Earth attracts the material of our bodies and keeps us firmly on the ground. Everything made of matter feels the pull of gravity. The Earth's attraction for matter also reaches out far into space, well beyond the moon. It pulls on everything that is there, although the force gets less as distance increases.

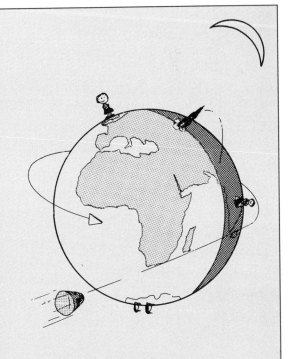

7 Here are some of the things that the force of gravity does:
   (a) It keeps us on the Earth's surface.
   (b) It holds the moon in orbit.
   (c) It captures returning spacecraft.
   (d) It brings down things we throw into the sky.
   If there were no pull of gravity, what would happen in cases (a) to (d)?

8 Invent some uses for a material that is repelled by the Earth's gravity. (Such a substance has yet to be discovered.)

9 This table gives the masses of the planets (compared to Earth) and their pull of gravity on a 1 kg mass near their surface.

	Mass	Pull (N/kg)
Mercury	0.04	2.8
Venus	0.83	8.9
Earth	1.00	10
Mars	0.11	3.9
Jupiter	318	25
Saturn	95	10.9
Uranus	15	11.0
Neptune	17	10.6
Pluto	0.06	2.8

   (a) On which planet would you feel heaviest?
   (b) Rewrite the table but in order of the pull of gravity of each planet.

Weight

The Earth pulls an object with a force that is called the weight of the object. As weight is a force, it is measured in newtons. If there were no Earth to pull the object, it would have no weight, but it would still have mass.

The Earth pulls a load of mass 100 g with a force of very nearly 1 newton. So the weight of the load is 1 newton on Earth. On the moon, the load would have less weight (about $\frac{1}{6}$ newton) because the moon pulls with less force than the Earth. (The load would still have a mass of 100 g.) In deep space the load would have zero weight although its mass would still be 100 g.

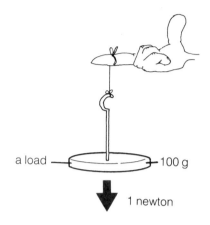

Feeling a pull of 1 newton

10 What is the weight on Earth of a 1 kg hanger?

11 A baby has a mass of 4 kg. What would be the weight of the baby on the Earth, on the moon and in deep space? Choose from these values: 6 N, 0 N, 40 N, 400 N. Copy and complete the table.

	Mass of baby	Weight of baby
On Earth		
On moon		
In deep space		

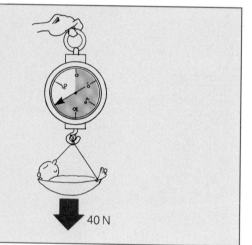

EXPERIMENT

Use scales to measure your weight in newtons.

From your weight work out your mass in kilograms.

12 A person found the Earth pulled her with a force of 500 N (her weight). Copy and complete this table for her. Use the information in question 9.

	Mass	Weight
On Earth		
On Venus		
In deep space		

newton scales

Weight watchers

'Weight' is a word that should be used with care. In everyday life we often use it to mean mass. In science weight is always the force of gravity on a body and is measured in newtons.

ordinary person   boffin

The (sometimes) troublesome force of friction

Feeling friction

If you rub a finger slowly across the back of your hand, you can feel friction pulling at your skin. Friction pushes the skin of the hand into wrinkles and drags at the finger as it moves along. Both skin surfaces feel the effect of the force as they slide over each other.

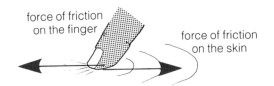

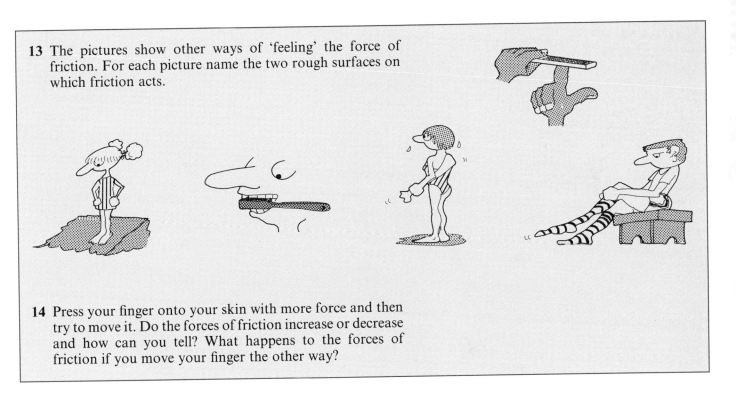

13 The pictures show other ways of 'feeling' the force of friction. For each picture name the two rough surfaces on which friction acts.

14 Press your finger onto your skin with more force and then try to move it. Do the forces of friction increase or decrease and how can you tell? What happens to the forces of friction if you move your finger the other way?

Good news and bad news

When friction acts, it can be useful or it can be a nuisance. Friction makes it possible for us to ride along on a bicycle but it also makes it harder to pedal.

15 This table describes forces of friction that act when we ride a bike. Copy the table, number each force and say whether it is useful or a nuisance.

Number	Force	Useful or nuisance
	friction on the tyres from the road	
	friction on the rider from the passing air	
	friction on the wheel rim from the brake blocks	
	friction on the feet from the pedals	
	friction in the wheel bearings	

16 The examples show the forces of friction in action. For each example:
(a) Name two surfaces between which friction acts.
(b) Say what the force of friction is doing.
(c) Say what would happen if the friction force suddenly disappeared.

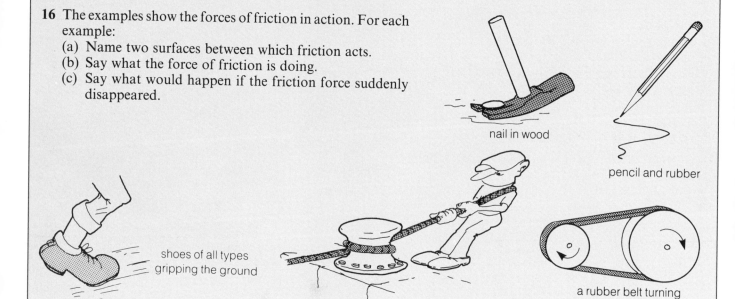

nail in wood

pencil and rubber

shoes of all types gripping the ground

a rubber belt turning a pulley

Experiments to reduce friction

As the Earth moves in space, it is almost completely free from friction. It spins without slowing down and, even without a push, will keep on turning for a long time to come. On the Earth, friction acts on everything that moves and prevents this sort of perpetual motion. To get some idea of what motion would be like without friction, try these simple ways of reducing it.

1. Air cushions
Blow gently under a sheet of paper that is lying on the bench. The paper will lift up and float away on a cushion of air.

2. Polystyrene beads
Scatter a handful of these small round plastic beads on a tray. Put a shoe on the beads and notice how freely it moves. The beads act like small wheels so that the shoe can roll along instead of having to slide.

3. Banana skin
Does a banana skin really reduce friction between a shoe and the ground? Try it and find out.

4. Skid-pads
Put a pool of oil onto a wooden board and release a spring driven toy car on the oil. Watch what happens to the car's motion when the friction on its wheels is reduced. Let the car drive through the patch of oil and study the way it skids when friction suddenly disappears.

More grip, less slip
We learn as children that friction can be increased by pressing surfaces together with more force. This method of increasing the force of friction has some useful applications. Some are shown below.

17 Write a sentence for each picture saying how the force of friction has been increased. Say also what the increased force of friction does in each case.

18 Knots and nails work because of the enormous forces of friction that can be produced when surfaces are pressed together. Copy these sentences about knots and nails and fill in the missing words:

The nail
When a nail is hammered ____ wood, it makes – tight-fitting hole for itself. The wood near ___ nail is compressed ___ pushes back on the nail ____ great force. Any attempt to move the nail must ____ overcome a greatly increased _____ of friction.

The knot
A knot holds two pieces __ rope together __ friction. When the ends of the ____ are pulled, the knot tightens ___ presses the two strands tightly _____. This increases friction _____ them, which makes it difficult for ___ ropes to slide apart.

Friction opposes motion

The friction on a body will always oppose any force that tries to move the body or that keeps it moving. When surfaces slide, friction causes wear and converts useful energy into heat. But when surfaces roll, friction prevents slip and moves the rolling part along.

Wear and heat are produced by friction when surfaces slide across each other.

When wheels roll, there is no slip. Friction forces the wheel one way and the road the other

19 The picture shows a person trying to slide a heavy box along the ground. The free-body diagrams show the main forces acting on the person, the box and the ground.

(a) Copy the table and put the correct number against the description of each force.

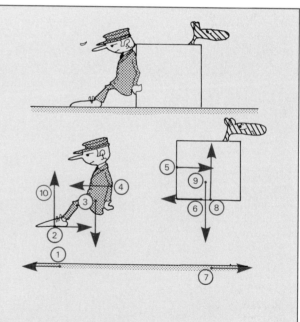

Number	Force
	friction of feet on the ground
	friction of ground on feet
	push of the person on the box
	reaction of box on person
	friction of the box on the ground
	friction of the ground on the box
	weight of box
	reaction of ground on box
	weight of person
	reaction of person on ground

(b) Which forces are equal in size?
(c) Which forces are action/reaction pairs?

20 An inventor decided that her locomotive was too sluggish. She reduced friction by fitting the wheels with ball-bearings, oiling the moving parts and streamlining the shape. Name three differences that she would notice when she runs her slick machine.

The friction of the air

We have all felt the force of the wind on our faces and how air resists when we move through it. There is friction whenever air meets a surface. Usually air resistance is a nuisance and objects designed for speed are streamlined to reduce the drag of the air. At high speeds most of the force resisting motion is due to the friction of passing air.

Sometimes objects are shaped to increase their wind resistance and make use of the friction of the air.

21 Give examples of
   (a) objects designed to let air flow over them smoothly,
   (b) examples of shapes that are designed to have a large air resistance.

Terminal speed

When we try to accelerate through the air, its resistance increases rapidly as our speed rises. The drag of the air can become so large that it balances the force that drives us on. When drag and driving force balance like this, there is no further acceleration and we travel on at constant speed. This top speed in air is called **terminal speed**.

22 These figures give the speed and times for a sky-diver as she free-falls from an aeroplane through still air.

Speed (m/s)	0	23.6	37.9	46.6	51.8	55.1	57.0	58.1	58.9	59.3	59.6
Time (s)	0	1	2	3	4	5	6	7	8	9	10

(a) Plot a speed/time graph of her motion and use it to find (i) her terminal speed, (ii) the time it took her to reach terminal speed. Sketch on the same graph how a speed/time graph would look for a person with greater air resistance.

(b) Draw a free-body diagram of the sky-diver showing the forces acting at terminal speed. What is the net force acting at this speed?

(c) Which of these phrases best describes the acceleration before terminal speed is reached: steady, increasing, decreasing?

2 Measuring forces

It is important to be able to measure the size of a force. One way to do this is with a force-meter or newton-meter. The unit used to measure force is the newton.

Feeling a force of 1 newton
The Earth pulls a load of mass 100 g with a force of very nearly 1 newton.

Tie a thread to a 100 g mass and hold onto the other end. This will give you a feel of a 1 newton force.

> 1 Which of these forces do you think are about 1 newton in strength: lifting an apple, kicking a ball, brushing your teeth, blowing away crumbs, stroking the cat, cracking an egg, pulling on socks, lifting a chair. Make a list of other forces that are about 1 newton in strength.

EXPERIMENT. A home-made force-meter
Take a strip of stiff card, about 30 cm long and narrow enough to fit inside an 'expendable' spring. Fix the top of the spring to the card with a split-pin and tie a string to the bottom end.

When a force pulls on the string, the spring stretches and covers more of the card.

How to find the newton marks
Hang a 100 g mass on your force-meter. This will pull with a force of 1 newton and stretch the spring. Mark the card at the bottom of the spring; this is the 1 newton mark. Repeat with extra 100 g masses to find the other newton marks.

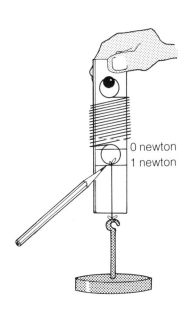

EXPERIMENT. Measuring some interesting forces
You will have to be inventive to measure some of the forces listed below. You may also need to use more than one force-meter at a time.

Measure the force needed to: lift a shoe, pick up a book, pick up a cup of tea, to open a door, to break a match stick, to break a biscuit, to pull on a sock, to double the length of an elastic band, to stretch old tights by 10 cm.

Make a table of your results and plot them into a bar chart.

2 This graph shows the length of a piece of rubber when stretched with different forces.

(a) When the rubber was tied to a door knob it stretched by 60 cm as the door was opened. What force was used to open the door.

(b) How long would the rubber be when pulled with a force of 200 N?

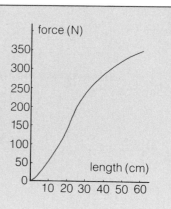

Spring balances

Your home-made force-meter is a simple spring balance.

A spring balance uses the extension of a spring to measure weight or pulling force. It should be marked in newtons. Sometimes spring balances have gears to change the stretch of the spring into the movement of a pointer. Your bathroom scales are spring balances.

  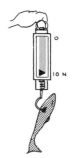

Measuring forces with a spring balance

EXPERIMENT. Measuring the force of your muscles

Use newton scales (bathroom scales marked in newtons) to measure the greatest force you can make with some of your muscles.

Put your results into a table like this:

	Force
Biceps	
Triceps	
Fingers	
Pectorals	
Thigh	

Plot a bar chart of these results.

Measuring the stretch

You may have found that the newton marks on your force-meter were evenly spaced. Experiment with other materials and measure how much they stretch when loaded.

EXPERIMENT. Stretching materials

Choose a material such as rubber, plastic or a spring that can be stretched easily. Use the weight of a metal hanger to pull the material and stretch it. Measure the amount the material stretches (its extension) for each weight you add. Put your results into a neat table. Collect results for a number of different materials. Plot your results on a graph with axes like these. The graph will show how the material behaved when it was stretched.

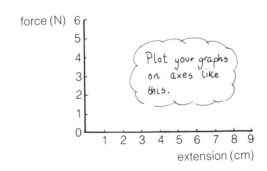

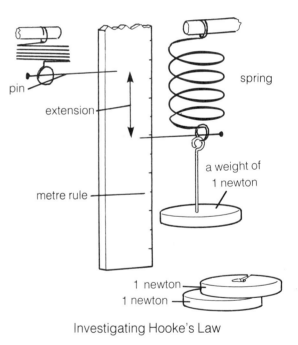

Investigating Hooke's Law

Hooke's law

This law says that the extension of a material is proportional to the force (i.e. if you double the force, you double the extension). If a material follows Hooke's law, a graph of force and extension will be a straight line through the origin.

Do the materials in your experiment follow Hooke's law for all or part of the stretch you gave them?

Elastic limit

Many materials only follow Hooke's law for part of their extension. After they have been stretched to a certain limit, the force and extension are no longer proportional. The graph of force and extension is no longer straight from that point and when the force is removed from the material it does not go back to its starting length. This limit is called the **elastic limit**.

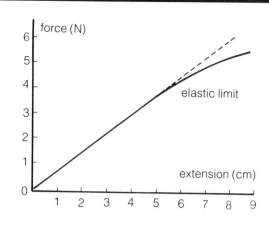

3 These results were obtained from a stretching experiment.

Force on the spring (N)	0	1	2	3	4	5	6
Extension (cm)	0	1.5	3.0	4.5	6.0	7.5	9.0

Force on the material (N)	0	1	2	3	4	5	6
Extension (cm)	0	3.5	7.5	11.5	15.5	18.5	20

Plot graphs of force against extension and mark any regions that follow Hooke's law.

4 Here are some examples of springy objects in common use: car springs, catapult elastic, elastic bands, elastic in clothes, mattress springs, chest expanders.

For each one write a sentence explaining what the force that the spring exerts is used to do.

5 A force of 20 N stretches a 'chest expander' by 0.5 m. How much force must a man use to stretch it by 1.5 m if the spring obeys Hooke's law? Is this force exerted by each arm, or do they share the force between them?

6 One of these arrangements of chest expanders below is four times harder to stretch (through the same distance) than one of the others. Which one is hardest to stretch and which one easiest? Can you explain why one needs four times more force than the easiest one?

A

B

C

3 Forces in balance

The turning effect of a force

We often use force to turn things.

The objects must be able to turn about an axis and the force must be made to act some distance from that axis.

> **1** (a) Draw simplified diagrams of the pictures and add force arrows to show where the force acts.
> (b) Mark the axis with an X and show which way the object turns.

EXPERIMENT. Increasing the turning effect

Hang a weight on a rod and move it further and further from the hand (see diagrams). Try this with a larger weight. It is clear that the 'turning effect' depends on the **weight** and the **distance** from the weight to the hand.

The turning effect of a force is called its **moment** and is calculated by multiplying the force by the perpendicular distance from where the force acts to the turning point.

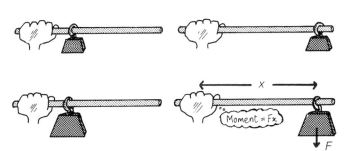

Increasing the turning effect of forces

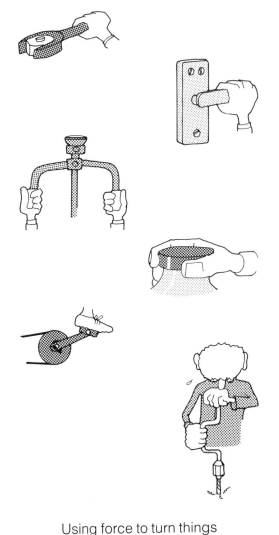

Using force to turn things

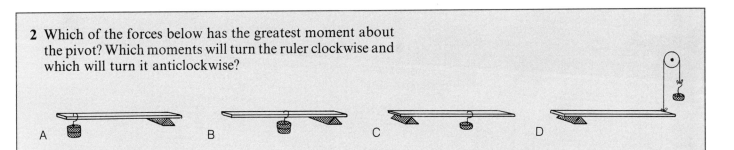

> **2** Which of the forces below has the greatest moment about the pivot? Which moments will turn the ruler clockwise and which will turn it anticlockwise?

3 Look at the spanner being used to turn a stubborn nut. Explain two ways of increasing the turning effects of the force on the nut.

Balanced moments

The weights of the little girl and her grandad have the same turning effect on the plank, but try to turn it in different directions. The moments of their two forces cancel and the plank doesn't turn at all. For the balanced plank, the moment trying to turn it clockwise equals the moment trying to turn it anticlockwise.

If several forces act on an object, then the moments trying to turn it clockwise can be added to give the total clockwise moment. This will then equal the total anticlockwise moment if the object is balanced.

So for any balanced object:

the sum of the clockwise moments =
the sum of the anticlockwise moments

This idea is known as the law of moments.

4 Say in the boxes whether the quantities shown are the *same* or *different* for the little girl and her grandad.

Weight	Distance to the log	Moment	Direction of turning effect

EXPERIMENT. Balancing a beam

You can test the law of moments by using a beam that is marked at regular intervals and a set of equal weights.

First make sure the beam balances on its centre line. Then place weights diagonally on the marks until the beam balances.

Each time you get it to balance, work out the moments of the weights on each side (the weight × its distance to the centre). You can then check to see whether the total clockwise moment always equals the total anticlockwise moment for a balanced beam.

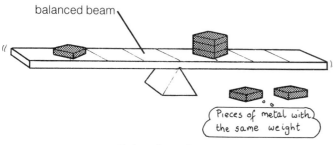

Balancing a beam

Stability and balance

We learn to balance things, including ourselves, at a very early age. We 'know' that a bottle of milk can be knocked over more easily than a loaf of bread, but what makes some objects more stable than others?

EXPERIMENT. Find the balancing point of a sheet of card

Take a sheet of card and hang it on a pin clamped in a stand. Make sure the card can swing freely. Make a 'plumb line' from a heavy nut and length of thread and hand it from the pin. Mark the line of the thread on the card. Do this twice more, hanging the card from different holes. You should find that the three lines cross at one point, and that the card will balance on a pin at that point. The whole weight of the card acts at that balancing point. It is called the **centre of gravity** of the card.

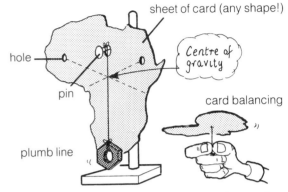

Finding the centre of gravity of a card

Stability and centre of gravity

A nut hanging on a thread is **stable**. If it is moved, it swings back to where it was. Note that its centre of gravity **rises** when it is moved. A balanced pencil is **unstable**. If it is moved its centre of gravity falls and it keeps on falling. A ball on level ground is in a **neutral** position. If it is moved, its centre of gravity does not rise or fall. It will stay where it is put.

 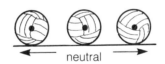 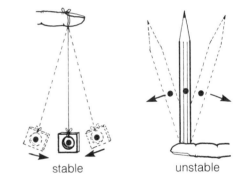

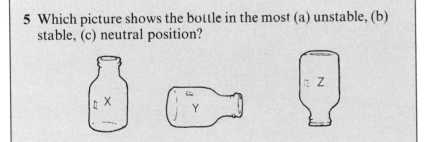

5 Which picture shows the bottle in the most (a) unstable, (b) stable, (c) neutral position?

An amazing balancing feat – balancing a pencil on a pin

It is impossible to balance a pencil on a pin unless you lower its centre of gravity. You can do this with 'sausages' of plasticine as shown. It will now balance in a stable position because the centre of gravity rises when it is moved.

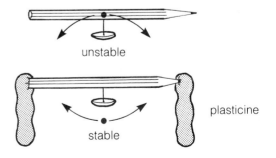

19

EXPERIMENT. Toppling over

Pin a plumb line to the centre of a heavy block of wood and tilt the block until it topples over. You will see that the block tips over when the plumb line passes the corner of the block. An object topples when its centre of gravity lies vertically outside of its base. This happens most easily when the centre of gravity is high and the length of the base is small.

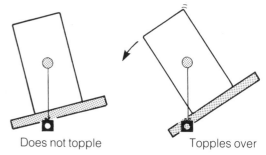
Does not topple   Topples over

6 What has the designer done to make these objects difficult to tip over?

  (a) A glass. Why it is less stable when it is full of liquid?

  (b) A sailing boat. Why is a boat less stable when people stand up?

  (c) A netball post on a stand. Why is it more likely to tip over if you climb up it?

  (d) A table lamp. Suggest a good material for the base.

7 Give two reasons why racing cars are difficult to tip over.

8 What advantage is it to a boxer, to crouch low and stand with his feet apart?

9 Why are steps rather easy to tip over sideways?

10 Why is a stool easier to tip over than a deck-chair?

11 On which face would you lay a match box to make it as stable as possible? Explain why this position is best.

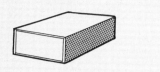

Further questions of balance

Use the law of moments to answer these questions.

12 The plank is uniform, weighs 200 N and is 3 m long. The painter weighs 600 N and he has put the tressels 0.5 m from the ends of the plank.
   (a) Draw a free-body diagram of the plank showing the four forces acting on it.
   (b) Draw a free-body diagram of the plank when the painter stands on the left-hand end.
   (c) Take moments and say whether the painter can walk to the end of the plank without it tipping up.

Example

Problem
Two strong ladies carry a log with two acrobats on their shoulders. The weights and positions of the people are marked on the picture. Work out the weight that each lady has to carry on her shoulders.

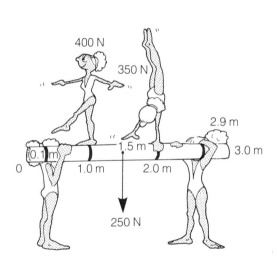

Solution
A free-body diagram for the plank looks like this. It doesn't move or rotate. P and Q are the pushes of the ladies on the log.

Work out the moments about P.

$$\text{Clockwise} = 400 \times 0.9 + 250 \times 1.4 + 350 \times 1.9$$
$$= 1375$$
$$\text{Anticlockwise} = Q \times 2.8$$
$$\text{So } 2.8\, Q = 1375$$
$$Q = 491 \text{ N}$$

The forces acting down add up to 1000 N.
The forces acting up are $P+Q$.
The log doesn't move up or down so the upward and downward forces are equal.

$$P + Q = 1000$$
$$P = 1000 - 491$$
$$P = 509 \text{ N}$$

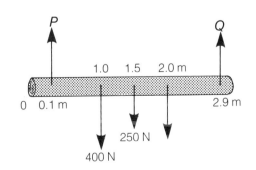

13 Work out the weight each lady carries if the acrobats change places.

Forces check list

After studying the section on Forces you should be able to:

- name different types of force
- know that force is measured in newtons
- know some effects of the force of gravity
- know that the force of gravity varies from place to place
- know the difference between mass and weight
- know how to draw force vectors
- know how to draw free-body diagrams
- know that action and reaction forces occur in pairs
- know Newton's third law of motion
- know some useful effects of friction and some examples of when it is a nuisance
- know that friction opposes motion
- know how friction can be reduced
- know how to increase friction
- know about air resistance and terminal velocity
- know how to measure force with a newton-meter
- know how to find out if materials follow Hooke's law
- know how to find the elastic limit
- know how to calculate the moment of a force
- know how to test the law of moments
- know how to find the centre of gravity of an object
- know and identify the types of stability

Revision quiz

Use these questions to help you revise the section on Forces.

- What can forces do to things?
  ... force can make things move, stop them moving, change their direction, change their shape and balance other forces to stop movement.

- Name some types of force.
  ... pushes, pulls, spring forces, gravity and friction (see also magnetic and electric on pp. 246 and 180).

- What is the force of gravity?
  ... it is the pull of the Earth's mass on a body.

- What affects the force of gravity?
  ... the mass of a planet and the distance to its centre of gravity.

- What is weight?
  ... weight is the pull of gravity on a body.

- What is the mass of a body?
  ... mass is a measure of the amount of material in a body or its inertia to movement.

- What are the units of mass and weight?
  ... mass is measured in kilograms and weight in newtons.

- How is force represented by a vector?
  ... the length of the vector gives the size of the force, the arrow gives the direction of the force and the force acts at the starting point of the vector.

- What is a free-body diagram?
  ... it is a diagram of part of a body showing all the forces acting on it.

- What is a reaction force?
  ... it is the force that acts back on an object when it exerts a force on another object.

- What is Newton's third law of motion?
  ... action and reaction are equal, opposite and act on different bodies.

- Which way does friction act?
  ... friction always acts to oppose motion and changes work into heat energy.

- How can friction be reduced
  ... between surfaces?
  ... between a body and air or liquid?
  ... by using air cushions, oil or ball-bearings.
  ... by streamlining or moving slower.

- How can friction be increased?
  ... by pressing the surfaces together with more force.

- What is a spring balance?
  ... it is a machine that uses a stretched spring to measure force.

- What is Hooke's law?
  ... the force on a material and its extension are proportional.

- What is the elastic limit?
  ... it is the extension to which Hooke's law is obeyed.

- On what does the turning effect of a force depend?
  ... on the size of the force and the distance to the axis.

- What is the moment of a force about a point?
  Moment = the force × the perpendicular distance to the turning point.

- What is the law of moments?
  For a balanced object, the total clockwise moments = the total anticlockwise moments.

- Where is the centre of gravity of an object?
  ... it is the point where the whole weight of the object can be thought to act.

- When does an object topple over?
  ... when a vertical line from its centre of gravity lies outside its base.

Examination questions

1 This question is about stretching a spiral spring.
   A loaded spring is mounted vertically as shown in the diagram. h is the height of the bottom of the load above the bench.
   (a) Describe how you would use a metre rule to measure h. Include the precautions you would take to make your results as reliable as possible.
   (b) A student measured values of h for several different values of load. The measurements are shown in the table.

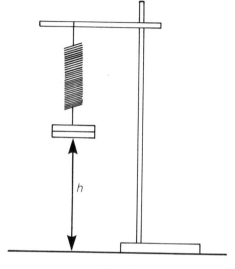

Fig. 1

Load in N	Height h in mm
1	184
2	172
3	162
4	150
5	141

   (i) Plot a graph of h (y-axis) against load (x-axis) on graph paper.
   (ii) Draw the best straight line.
   (iii) Use your graph to find the load which gives a value of h of 180 mm.
   (iv) Use your graph to find the value of h at a load of 1.50 N.
   (v) Use your graph to help you calculate the CHANGE IN LOAD which gives a CHANGE IN h of 1.00 mm.

   (*SEG*)

2 The diagram shows a jet aircraft which has a mass of 50 000 kg.

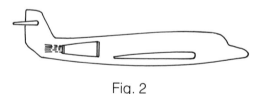

Fig. 2

   (a) What is the weight of the aircraft?
       Weight is a force. Copy the diagram and draw an arrow, labelled W, to show the direction of this force.
   (b) What is the value of the upward force (lift) on the aircraft just as it is leaving the ground?
       On your diagram draw another arrow, labelled L, to show the direction of this force.
   (c) The aircraft engines are at full thrust as it accelerates along the runway and takes off. Where do you think the acceleration is greatest? Give a reason for your answer.

   (*MEG*)

3 A piece of elastic used in a catapult has a force–extension graph as shown. When the elastic was stretched to the maximum extension shown on the graph, the energy stored in the elastic was 25 J. Make a copy of the graph.
   (a) Does the elastic obey Hooke's law over the range of values shown? Justify your answer.
   (b) When the stone was released from the catapult, 80% of the energy stored in the elastic was given to the stone as kinetic energy. The stone, which had a mass of 100 g, took 0.2 s to leave the catapult. (See p. 85.)
   What was the maximum speed of the stone? Explain your working.
   (c) The original piece of elastic was replaced by a piece twice as long.
      (i) On your graph sketch as accurately as you can the force–extension graph that you would expect to obtain for this new piece of elastic.
      (ii) Explain the shape of the graph you have drawn.
   (*MEG*)

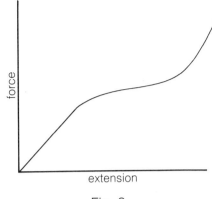

Fig. 3

4 A rectangular block measures 8 cm by 5 cm by 4 cm, and has a mass of 1.25 kg.
   (a) (i) If the gravitational field strength is 10 N/kg, what is the weight of the block?
      (ii) What is the area of the smallest face of the block?
      (iii) What pressure (in N/cm^2) will the block exert when it is resting on a table on its smallest face? (See p. 53.)
      (iv) What is the least pressure the block could exert on the table?
   (b) (i) What is the volume of the block?
      (ii) Calculate the density of the material from which the block is made. (See p. 299.)
   (*SEG*)

5 When ball bearings are manufactured it is important that they have the correct mass. The diagram below illustrates a device for doing this.
   A uniform beam AB is pivoted at its centre 0. It has a weight W of 1 N placed at end B, and a container of negligible weight placed at X. When the container is filled with 100 ball bearings, the beam should balance if the ball bearings have the correct mass. Assume that a mass of 1 kg has a weight of 10 N.
   (a) What is the moment of the weight W about the pivot?
   (b) Calculate the weight (in newtons) of the ball bearings in the container at X.
   (c) What is therefore the average mass (in grams) of **one** ball bearing?
   (d) If the container was filled with 100 ball bearings of identical size but lower density, would the end A of the beam tip up or down? Give a clear reason for your answer.
   (*NISEC*)

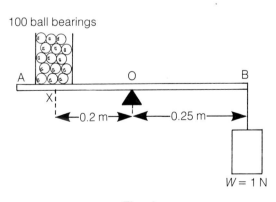

Fig. 4

4 Measuring motion

Timers

We are always measuring time intervals. Here are some examples of timers.

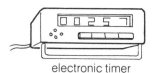
electronic timer

watch

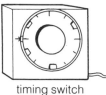
timing switch

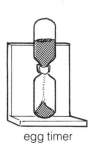
egg timer

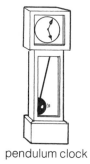
pendulum clock

1 Copy this table and put in a √ in the box if the timer CAN measure that time (interval) and an X if it cannot.

Timer	Hours	Minutes	Seconds	$\frac{1}{10}$th of seconds	$\frac{1}{100}$th of seconds
Electronic timer					
Egg timer					
Stopwatch					
Timing switch					
Pendulum clock					

2 Give an example of a use for each timer.

The ticker-timer

This very useful laboratory timer prints dots on paper tape that passes under its vibrating arm. It always prints 50 dots each second, so by counting the dots we can work out time.

25 dots would take 25/50 second (0.5 s) and 10 dots 10/50 seconds (0.2 s) for example.

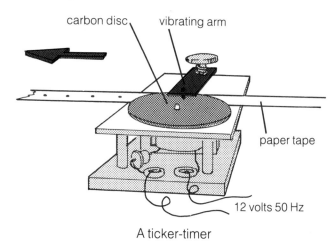

A ticker-timer

3 For each tape work out how long it took to print the dots.

4 Which tape was pulled through the timer fastest? Explain your answer.

Note that the first dot is printed when the time is 0 s and so should be counted as dot 0.

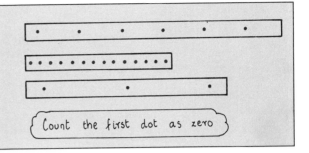

Count the first dot as zero

Distance and time

Recording motion

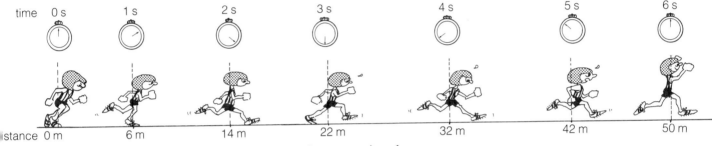

A record of the motion of this runner can be kept by measuring the distance she has covered at different times.

Time (s)	0	1	2	3	4	5	6
Distance (m)	0	6	14	22	32	42	50

This set of numbers tells us where the runner was after each second but not where she was at any other time. We can predict where she was at in-between times by plotting the numbers on a distance/time graph. If the points are joined with a smooth line, we can read off where the runner probably was at any time. A distance/time graph is a useful way of recording motion

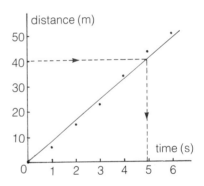

A distance/time graph of the sprint

5 Plot a distance/time graph of the motion of the runner. and find out:
   (a) the time when the sprinter was 20 m and 40 m from the start;
   (b) how far the sprinter had travelled in 3.5 s

6 Why can't we be absolutely sure of the runner's position at in-between times?

7 Which distance/time graphs show the motion of the following 'objects':
   (a) a snooker ball travelling across a table;
   (b) a cat asleep in a chair;
   (c) a ball thrown up into the air?

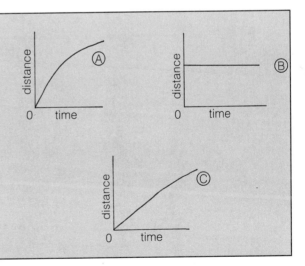

EXPERIMENT. Studying the motion of a toy car

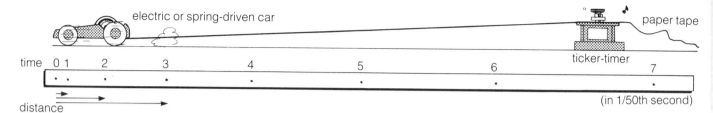

Obtain distance and time figures for the motion of the car by making it pull a paper tape through a ticker-timer. The timer prints dots on the tape every $\frac{1}{50}$ second at the distance covered by the car from the start of the motion. Measure from the first dot to all the other dots in turn to get the distances travelled at intervals of $\frac{1}{50}$ second. Use your results to plot a distance/time graph of the motion.

8 Continue the measurements taken on the tape above and plot a distance/time graph for the toy car. What is the total distance covered by the car during the time shown by the tape? What is the average speed of the car during this time?

9 Copy the table and write in the boxes which regions of this distance/time graph show the object:

stationary at home	
moving slowly	
moving quickly	
stationary away from home	

(Note that if the line is steep, the object must have moved a long way in a short time and its speed must have been high.)

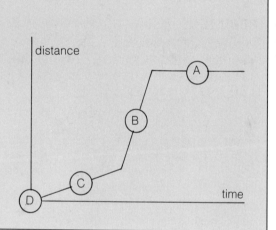

Speed

The speed of a body is the distance it moves in one second (or one hour). Speed can be worked out from this equation:

$$\text{speed} = \frac{\text{distance}}{\text{time}}$$

Speed is measured in miles per hour (mph)
    or kilometres per hour   (km/h)
    or metres per second   (m/s)
    or centimetres per second   (cm/s)

Car speedometer

A high-speed train

10 Copy this table and work out the numbers that go in the spaces.

Speed (m/s)	Distance (m)	Time (s)
	100	5
5		10
8	96	

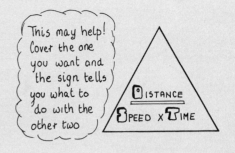

This may help! Cover the one you want and the sign tells you what to do with the other two

Distance / Speed × Time

11 Copy this table and put these speeds into the correct spaces: 8 km/h, 25 km/h, 120 km/h, 1000 km/h, 2000 km/h, 24 000 km/h.

Car (top speed)	Bike	Passenger jet	Walking	Concorde	Satellite

EXPERIMENT. Using a ticker-timer to measure speed

Pull a paper tape through a ticker-timer at a steady speed. It takes 1 s for the timer to print 50 dots, so count 50 dots and measure the distance they cover. This is the distance the tape has moved in 1 s and so is the speed of the tape.

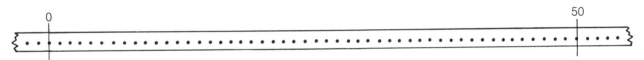

Example
The distance covered by these 50 dots is 15 cm. The speed of the tape is therefore 15 cm/s.

12 Work out the speeds of these tapes, shown full size.

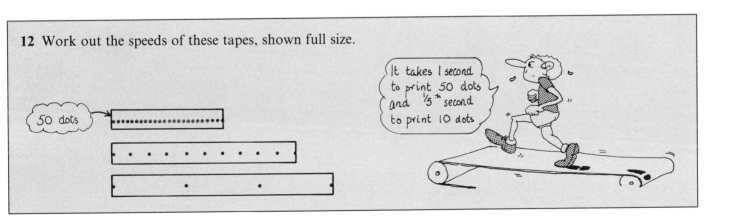

It takes 1 second to print 50 dots and 1/5 th second to print 10 dots

Speed and time

On any journey, speed will change as time goes by and as we travel it is important to know how fast we are going. Car speedometers give instant speed readings and the pictures show how the speed of a car changed as it moved along a short street.

clock

speedometer

Time (s)	0	1	2	3	4	5	6	7
Speed (m/s)	0	6	12	18	18	18	9	0

The motion can be visualized more easily by plotting the speed and time figures on a graph.

Speed/time graphs show:
Ⓐ when an object is not moving
Ⓑ when an object's speed is rising
Ⓒ when an object's speed is steady
Ⓓ when the object's speed is dropping

How long was the journey

The area under a speed/time graph is the total distance covered during the motion. This distance can be found by dividing the graph into sections and adding each of the areas together. Note that lengths and heights must be read off the axes using the appropriate scales.

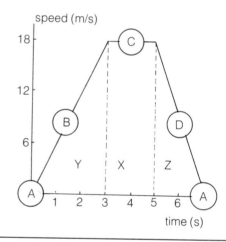

Proof that the distance travelled = the area under the speed/time graph.

Over section C: speed = 18; time = 2.
distance = 18 × 2 = 36 m.
This is also area X.

Over section B: average speed = 9;
time = 3.
distance = 9 × 3 = 27 m.
This is also area Y.

Over section D: average speed = 9;
time = 2.
distance = 9 × 2 × 18 m.
This is also area Z.

So total distance =
81 m = total area = X + Y + Z.

13 Calculate the distance travelled by the car above and the average speed of its journey along the street.

14 Use these speed and time figures of a journey to:
(a) plot a speed/time graph of the journey;
(b) find the total distance travelled;
(c) find the average speed during the journey.

Time (s)	0	1	2	3	4	5	6	7	8	9	10
Speed (m/s)	0	0	0	1	2	3	3	3	3	1.5	0

Put the following labels on your graph: 'stopped', 'steady speed', 'speed increasing', 'speed decreasing'.

Velocity

It would be no use directing the jogger in the forest by telling her to run at a speed of 3 m/s. She wouldn't know which direction to take. To find the ducks, for example, she must be told to run at 3 m/s east for 100 s. '3 m/s east' is called the traveller's **velocity**. Velocity is speed in a named direction and has two parts, **speed** and **direction**.

When the girl reaches the ducks, the distance and direction from the starting point (i.e. 300 m east) is called her **displacement**. Displacement is distance in a named direction.

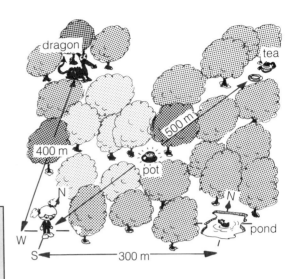

15 Copy this table and write in suitable velocities and times for the jogger to get to the following places in the forest:

Velocity		Time (s)	Destination
Speed (m/s)	Direction		
3	east	100	duck pond
			cup of tea
			cup of tea via the dragon
			pot of gold

Vectors and scalars

To describe the velocity of a body we must give its speed and its direction of travel. Quantities that have size and direction are called **vector** quantities and so velocity and displacement are vectors. (Vectors are often printed in bold type, e.g. **R**.) Time or temperature on the other hand can be described by just a number as they have no direction. These are examples of scalar quantities.

16 On a copy of this table tick or cross whether these quantities have size or direction and decide whether they are scalars or vectors.

	Wages	A push	Temperature	Time	Compass bearing	Velocity	Speed	Acceleration	A chess move
Size ?									
Direction ?									
Vector or scalar?									

Diagrams of vectors

Scalar quantities can be written just as a number (usually with a unit) but vector quantities cannot. They have to be shown by a line. The length of the line, drawn to scale, shows the size of the vector quantity and the direction of the line, its direction.

> **17** Which of these diagrams could represent each of the following vector quantities?
> (a) A diagonal chess move of two squares by a bishop.
> (b) A force of 6 N to the left.
> (c) A velocity of 3 m/s east.
> (d) A vertical jump of 20 cm.
> (e) The force of a 200 N weight resting on the ground.

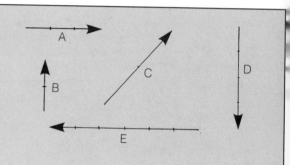

Adding velocities

Often an object can be given two velocities at once. A hovercraft, for example, can be given a velocity by its engine and another by the wind. The two velocities then 'add' to give the craft its actual velocity. When vectors are added their size and direction both affect the result. If the engine speed is 4 m/s and the wind speed 3 m/s, the craft's actual speed can work out to be 1 m/s or 7 m/s or 5 m/s depending on the wind direction.

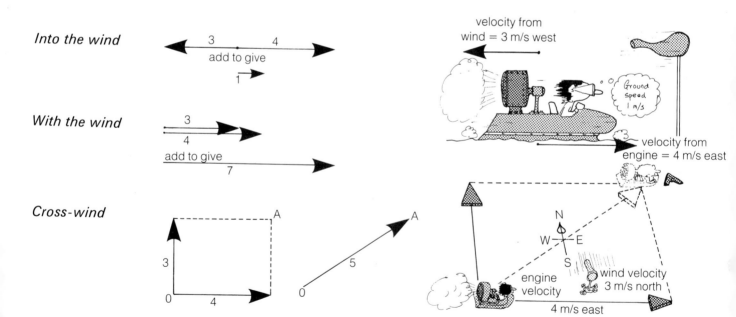

Flying in a cross-wind

In 1 s the engine of the hovercraft drives it 4 m east and at the same time the wind blows it 3 m north. The craft finishes up at A, 4 m east and 3 m north from its starting point. So the diagonal line OA is the craft's actual velocity vector.

How to find the sum of two vectors

R is the vector that you get if you add vectors **X** and **Y**.

The length of **R** can be calculated using Pythagoras' theorem:

$$R^2 = X^2 + Y^2$$

and the angle **R** makes with **X** from:

$$\tan \theta = \frac{Y}{X}$$

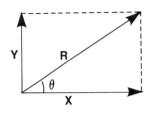

Example

If $X = 4$ m/s, $Y = 3$ m/s
then $R^2 = 4^2 + 3^2 = 25$
$R = 5$ m/s
$\tan \theta = \frac{3}{4} = 0.75$
$\theta = 36 \cdot 9°$

Vectors of all types can be added in this way.

---

18 Find the result of adding the following velocities:

30 m/s S	and	40 m/s W
5 m/s N	and	12 m/s E
15 m/s W	and	20 m/s N
15.6 m/s E	and	9 m/s S

In each case find the length (= speed) of the resultant vector and the angle it makes with north (its bearing).

Acceleration

If a tape is moved through the ticker-timer at a steady speed, the dots on it will be evenly spaced. If you cut the tape every 5 dots and stick the pieces side by side, they will all be about the same length.

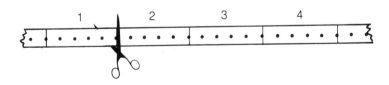

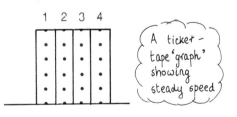

A ticker-tape 'graph' showing steady speed

EXPERIMENT

Pull a paper tape through a ticker-timer at an increasing speed. You could do this by sticking the tape to a trolley that can run down a sloping track. You will notice that the dots on the tape get further and further apart as the speed increases. Cut the tape every 5 dots as before and stick the pieces side by side (see next page). This time the lengths of the pieces increase, showing that the speed of the tape is increasing all the time. A body whose speed increases as time goes by is accelerating. (A body that goes slower and slower as time goes by has a negative acceleration.)

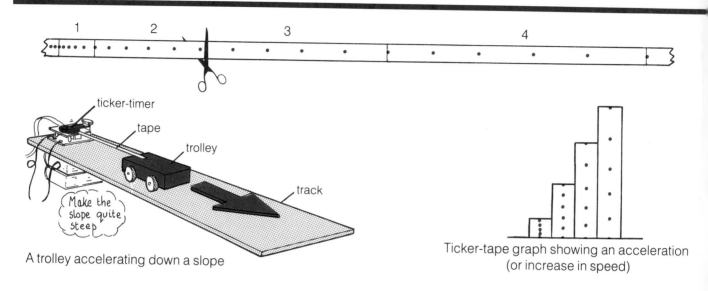

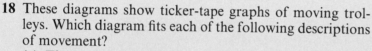

A trolley accelerating down a slope

Ticker-tape graph showing an acceleration (or increase in speed)

18 These diagrams show ticker-tape graphs of moving trolleys. Which diagram fits each of the following descriptions of movement?
(a) The trolley has a positive then negative acceleration.
(b) The trolley accelerates, travels steadily, then has a negative acceleration.
(c) The trolley accelerates, travels steadily, then accelerates again.

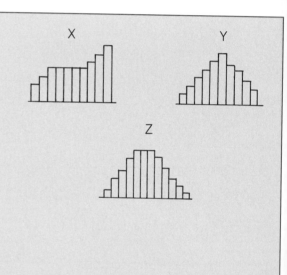

19 Draw a ticker-tape graph for a trolley that accelerates, travels at a steady speed, accelerates again and then has a negative acceleration.

20 A boy on a bicycle travels 300 m during one minute and 420 m during the following minute. Calculate his average speed:
(a) during the first minute,
(b) during the second minute,
(c) for the two-minute period.
(d) Was the boy's acceleration positive or negative during the two-minute period?

⊙ EXPERIMENT. Catapulted motion

Use a ticker-timer to get a speed/time graph for a trolley that is being accelerated by stretched elastic. Fix a long piece of elastic to one end of the trolley and paper tape to the other. Thread the tape through a ticker-timer, stretch the elastic, switch on and let go. As the elastic accelerates the trolley, the timer prints dots on the tape that record the distance travelled every 1/50 s.

Convert the tape into a speed/time graph by cutting it into lengths that are 5 dots long. These lengths represent the speed at 0.1 s intervals, so glue the pieces side by side to get the speed/time graph of the motion.

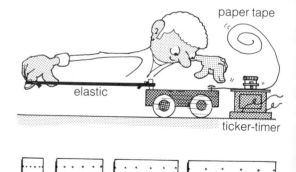

21 A student's speed/time graph for a catapulted trolley looked like this. Does it show that:
 (a) the acceleration of the trolley was constant;
 (b) the increase in speed got less as time passed;
 (c) the acceleration got less as the elastic shortened;
 (d) the trolley had a decreasing acceleration;
 (e) the trolley got faster and then began to slow down?
 Write down the three sentences that are correct.
 Explain why stretched elastic does not produce a constant acceleration.

22 An object accelerates whenever its velocity changes. This can be a change in its size (speed) or direction or both. Copy and complete this table about acceleration.

Velocity vector before	Velocity vector later	Does speed change?	Does direction change?	Is there an acceleration?
→	⟶			
→	↗			
⟶	→			
→	↗			
⟶	→			

Calculating acceleration

Rapid acceleration can often get us out of trouble and, on other occasions, slow acceleration is best.

Acceleration is measured by the change in velocity per second. For motion in a straight line it can be calculated from the formula:

acceleration = change in speed/time taken for the change

The unit used is m/s^2.

For example if the boxer moves his head from a speed of zero to 5 m/s in one second, the acceleration of his head is 5 m/s^2.

And if the lady moves the water pot from a speed of zero to 5 m/s in five seconds, the acceleration of the pot is 1 m/s^2.

Using fast and slow accelerations

23 Calculate the values that are missing in this table.

Speed at the start (m/s)	Speed at the end (m/s)	Change in speed (m/s)	Time from start to finish(s)	Acceleration (m/s^2)
0	15		0.5	
30	38		2	
20	0		4	
0			4	5

Three equations of motion

The acceleration formula can be used to work out three equations of motion. These can be used to calculate speeds, times and distances travelled during motion. The equations link:

- the start speed (u)
- the finish speed (v)
- the distance travelled (s)
- the acceleration (a)

They can only be used for steady acceleration in a straight line.

The equations are:

(1) $v = u + at$ if $u = 0$ $v = at$
(2) $s = ut + \tfrac{1}{2}at^2$ if $u = 0$ $s = \tfrac{1}{2}at^2$
(3) $v^2 = u^2 + 2as$ if $u = 0$ $v^2 = 2as$

The cartoon shows how these equations are worked out from the acceleration and speed formulae.

The bike changes speed from u to v.
This happens in time t.
So acceleration $a = (v-u)t$.
Rearranging $v = u + at$.
The average speed $= (v+u)/2$.
Time of travel $= t$.
So distance travelled $s = t(v+u)/2$.
Substituting for v: $s = ut + \tfrac{1}{2}at^2$.
Substituting for u: $v^2 = u^2 + 2as$.

Motion picture problems

Use the equations of motion to work out the unknowns (in red) in the following problems. A useful technique is to list the values you are given so that you can choose the best equation to calculate the unknown.

24 A skater moves off from rest with a steady acceleration of 4 m/s². What is her speed and distance travelled after 10 s?

25 An astronaut in deep space accelerates from 0 to 9 m/s in 6 s. What is his acceleration? He switches his jet off and drifts for a further 20 s. What is his speed then?

26 A snooker ball is given a speed of 8 m/s. It moves across the table with an acceleration of -1 m/s^2. How long does it take for the speed to drop to zero and how far will it have travelled?

27 A wheelchair athlete accelerates at 3 m/s^2. What is his speed after 5 s? He then accelerates at -0.5 m/s^2. How long will it take for his speed to reach zero and how far will he have travelled?

Sketch speed/time graphs for all four motions.

5 Force and motion

The way force affects motion can be summarized as follows:

(1) If there is no force (or no net force) there will be no acceleration. The body will stay still or continue to move at a steady speed. (This is known as Newton's first law of motion.)

(2) An unbalanced (or net) force will always produce an acceleration.

As an example of these ideas think about this brave skater. He moves on very smooth ice and has a rocket on his back to apply a force.

(1) No (net) force and no acceleration.

When there is no force there is no motion.

Where there is no net force there is no motion.

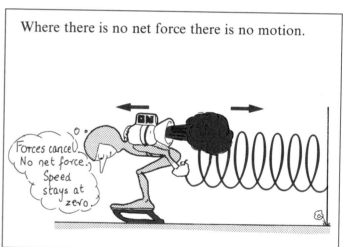

When there is no force and he is moving, the speed stays steady.

When there is no net force and he is moving, the speed stays steady.

(2) A net force always produces an acceleration.

A force in the direction of motion produces an acceleration that increases speed.

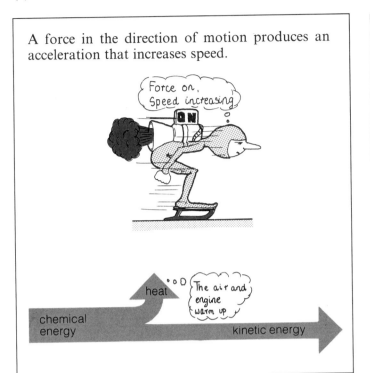

A force against the direction of motion produces an acceleration that decreases speed.

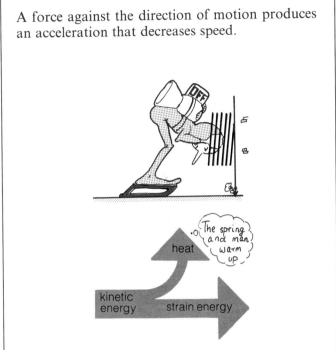

1 Copy this box and fill in what happens to the speed in each case.

	No force	No net force	Force acting
Still body	speed . . .	speed . . .	speed . . .
Moving body	speed . . .	speed . . .	speed . . . or . . .

2 In the cases where a force acts, energy changes take place. These are shown by the arrows. Describe these energy changes (see p. 75).

A car travelling at a steady speed along a level road is rather like the skater dragging an anchor through ice. The engine (through the wheels) exerts a forward force on the car in the same way as the rocket does on the skater. Wind resistance exerts a backwards force on the car, as does the drag of the anchor on the skater. These two forces are equal and balance, leaving no net force on the car. The car continues to move at a steady speed.

3 Copy and complete this table about the motion of a car.

Force and resistance	What happens to the car
Engine force greater than wind resistance	
Engine force less than wind resistance	
Engine force equal to wind resistance	

4 Sketch a speed/time graph for this brave skater.

Force and acceleration

The pull of gravity on the metal mass in the picture is felt by the finger.

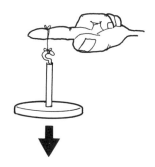

EXPERIMENT. Force and acceleration

Use this force to accelerate a trolley as shown in the picture. Fix a paper tape to the trolley and use a ticker-timer to print dots on the tape as it accelerates by. Cut the tape every 5 dots and stick the pieces side by side. This will give a ticker-tape graph that shows that the trolley is speeding up (accelerating). Do the experiment again but this time use the weight of two of the masses to move the trolley. Again cut and stick the tape to give a ticker-tape graph.

You will see from your results that the acceleration gets larger as the force gets larger. In fact, careful experiments have shown that doubling the force doubles the acceleration. Force and acceleration are proportional.

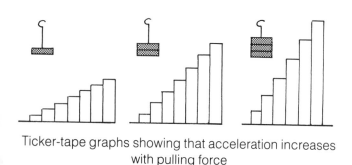

Ticker-tape graphs showing that acceleration increases with pulling force

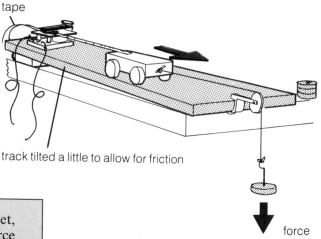

Using a force to accelerate a trolley

5 The following objects can accelerate: sprinter, car, bullet, skydiver. For each one say what it is that provides the force for acceleration.

Newton's second law of motion

This law states that:

the acceleration of an object is proportional to the force on it provided its mass doesn't change.

EXPERIMENT. To find if a trolley on a track follows Newton's second law

Use a ticker-timer to measure the acceleration of a trolley for different forces.

First check that the tilt of the track is just enough to balance the friction of the trolley as it pulls tape through the timer. Alter the angle until a small push makes the trolley move at a steady speed.

Then use your home-made newton-meter (see p. 13) to accelerate the trolley with a 1 newton force. Keep the force steady as the trolley moves and obtain a ticker-tape of the motion. Make sure that the tape is at least 50 dots long and then repeat for other forces.

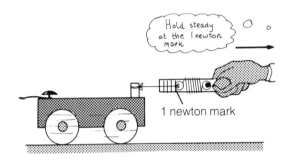

Calculating the acceleration

Measure the length (s) of the first 50 dots and use the equation $s = \frac{1}{2} at^2$ to calculate the acceleration (a).

50 dots take one second ($t = 1$) and so $a = 2s$. Put your results into a table like this:

Force	0	1	2	3	4	5
Acceleration (cm/s²)						

Plot a graph of acceleration against force. If the points lie close to a straight line that goes through the origin, it shows that the motion of your trolley follows Newton's second law. What is the conclusion from your results?

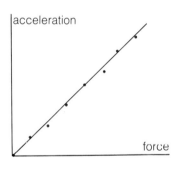

6 Calculate the acceleration from the tapes shown and plot a suitable graph to find out whether the motion follows Newton's second law.

ticker-tapes 50 dots long

An equation for Newton's second law

Careful experiments have confirmed that force is proportional to acceleration and that acceleration is inversely proportional to the mass of a body. This means that force is proportional to mass × acceleration.

The unit of mass is chosen so that the equation can be written as:

$$\boxed{\begin{array}{c}\text{force} = \text{mass} \times \text{acceleration} \\ F = ma\end{array}}$$

Units

mass is measured in kilograms (kg)
acceleration in metres/second^2 (m/s^2) force in newtons (N)

where 1 newton is the force/kilogram needed to produce an acceleration of 1 m/s^2.

More picture problems

The questions that follow have been laid out in a way that shows a useful technique for solving problems. Usually questions describe a situation in words, add some numerical data and ask you to calculate certain 'unknowns'. A useful first step is to draw a picture that includes the known quantities and to make a list of what is given and what is required. You can then choose an appropriate equation to work out the answer.

7 A rocket pushes a 75 kg skater with a force of 600 N. What is the acceleration produced and the speed after 8 s?

knowns and unknowns:
$F = 600$ N
$m = 75$ kg
$a = ?$
$v = ?$
$t = 8$ s

Equations:
$F = ma$
$v = at$

8 An invalid in a wheelchair (total mass 85 kg) pushes himself forwards with a force of 100 N. What is his acceleration and how far will he travel in 13 s?

$F = 100$ N
$m = 85$ kg
$a = ?$
$s = ?$
$t = 13$ s

$F = ma$
$s = \frac{1}{2}at^2$

9 What force must a cyclist use to reach a speed of 15 m/s in 5 s, if man and machine weigh 95 kg?

$F = ?$
$m = 95$ kg
$a = ?$
$v = 15$ m/s
$t = 5$ s

$F = ma$
$v = at$

The acceleration due to gravity

When you let go of an object above the ground, the pull of gravity makes it fall faster and faster. The pull of the Earth accelerates the object.

EXPERIMENT. The motion of a falling stone

Make a ticker-tape of the motion of a falling stone.

Fix a ticker-timer vertically and several metres above the floor. Hang the stone on the end of a long piece of ticker-tape paper. Thread the paper through the timer and pull the stone up to the top. Switch on the timer and then let the stone fall.

Several dots should be printed on the tape in the short time that it takes the stone to reach the ground. Look at the spacing of these dots and decide whether the stone is accelerating.

To show the motion more clearly make a ticker-tape graph by cutting the tape at every dot and sticking the pieces side by side. If the pieces get longer and longer, it shows that the stone is being accelerated by the pull of gravity.

10 This tape was made by a falling stone.
    (a) Which end was the stone fixed to?
    (b) How long did the stone take to fall?

The pull of gravity gives all objects the same acceleration

It is hard to believe, from everyday experience, that light objects fall with the same acceleration as heavy ones. This is because we usually watch objects fall through air and the air slows them down. A sheet of paper, for example, takes longer to reach the ground than a tennis ball released from the same height.

 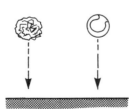

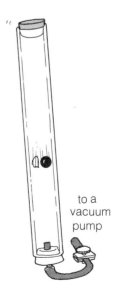

to a vacuum pump

If the effect of the air is reduced by crumpling the paper into a ball, then paper and ball do hit the ground together.

A better test is to put a ball and a piece of paper into a long sealed tube and to suck out the air. When the tube is up-ended, the ball and paper fall and reach the bottom exactly together.

11 Can the pull of gravity act through a vacuum? How could you prove that your answer is correct?

Free-fall

EXPERIMENT. Measuring the acceleration of gravity

The acceleration of an object when it free-falls to the Earth is called g. This experiment measures the value of g by finding the distance and time of free-fall of a steel ball. The picture shows apparatus that measures the time of fall automatically.

Use an electronic stopwatch and arrange contacts so that the ball starts the clock when it begins to fall and stops it when it reaches the ground. To make the 'start' contacts cut a large hole in the centre of a piece of thick card. Stick strips of aluminium on each side of the hole so that when the ball falls through it softly brushes both strips. Staple long wires to the strips and attach these to the make-to-start contacts on the clock. The ball should then start the clock when it falls through the hole.

The stop contacts are two squares of aluminium foil stuck on either side of a thick cardboard frame. When the ball lands on the top square make sure that it is pressed down onto the bottom square. Leads from these squares to the make-to-stop contacts should then stop the clock when the ball lands.

Arrange the 'start card' several metres above the 'stop card' and measure at least 10 readings of the time of fall. These will differ slightly due to small errors beyond our control. Some readings will be slightly higher than the true time and some lower, so take an average to find the best value (t). Also measure the exact distance of fall (s) between the contact plates. Then the acceleration of the ball can be calculated from the equation $s = \frac{1}{2}at^2$.

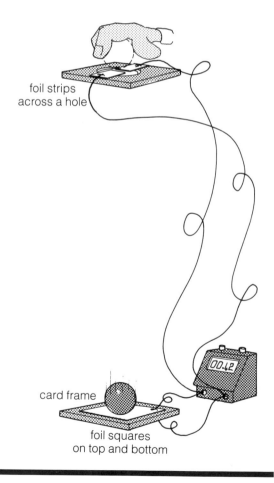

12 Use these figures for the time (in seconds) of fall of a steel ball, to calculate a value for *g*. The distance of fall was 1.8 m.

0.58, 0.59, 0.62, 0.61, 0.57, 0.61, 0.63, 0.59, 0.60, 0.60.

13 Why it is impossible to measure these times accurately with a hand-held stopwatch?

14 This piece of ticker-tape (shown one tenth of its actual size) was printed by allowing a heavy weight to pull it vertically through a ticker-timer. Use the information in the dots to calculate a value for the acceleration of gravity. Suggest why you might expect the value to be rather low. (50 dots take 1 s to print.)

Values of *g*

The size of *g* depends on the mass of the planet and the distance to its centre. On Earth, *g* varies slightly from place to place but is usually taken as 10 m/s² for simple calculations.

Some values of *g* on Earth

	North Pole	Equator	Britain	Top of Everest
g (m/s²)	9.83	9.78	9.81	9.77

g gets less as you move further from the centre of the Earth. The spin of the Earth also reduces *g*, especially at the equator.

15 Name three ways of giving a person an acceleration of more than one *g* on Earth.

16 Name three differences that you would notice if *g* were one tenth of its normal value on Earth.

Gravitational field strength

It is the gravitational pull of the Earth that gives bodies a free-fall acceleration of 10 m/s². The size of this pull can be calculated from $F = ma$. On a 1 kg mass the force must be 10 N (1 × 10); on a 10 kg body it must be 100 N (10 × 10). So the pull of the Earth on each kilogram close to its surface is 10 N. The force per kilogram (10 N/kg) is called the **Earth's gravitational field strength** (see p. 5).

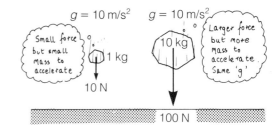

17 Explain why large and small masses have the same acceleration in the Earth's gravitational field.

18 Neglect the effect of air resistance in the problems that follow.

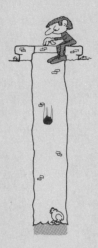
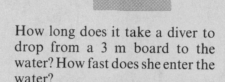

A stone takes 2.5 s to reach the bottom of a well. How deep is the well? How fast is the stone travelling at the bottom?

How long does it take a diver to drop from a 3 m board to the water? How fast does she enter the water?

How long does it take a coconut to fall from a 10 m tree. How fast does the nut hit the ground?

19 Calculate the distance and speed of an object at one-second intervals as it falls freely under gravity ($g = 10$ m/s^2).

Time (s)	0	1	2	3	4	5
Distance (m)	0					
Speed (m/s)	0					

The free-fall sensation

Earth-bound people do not experience free-fall for long and the sensation is often spoiled by the problem of landing.

To stand on firm ground, the legs have to be stiff enough to withstand body weight pressing from above and the reaction of the ground pressing from below.

The effort needed to support the body makes us conscious of its weight. When the ground is removed and we free-fall, the legs do not have to support the body and we feel weightless. But the body still has weight and is still attracted and accelerated to the Earth.

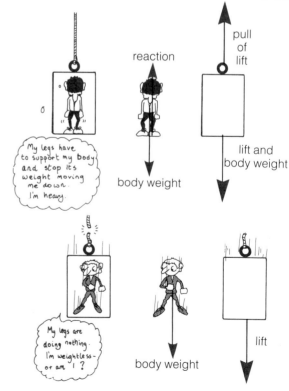

Humans in orbit

Spacemen who float around inside orbiting space stations are not weightless, but are free-falling towards the Earth. They do not appear to be falling because their craft is falling too, and they do not have to press against its sides to support their weight. The reason a free-falling craft does not hit the Earth is because it has enough orbital speed to keep it in orbit. The craft can free-fall continuously without getting any closer to the Earth's surface and the men inside can free-fall without ever landing.

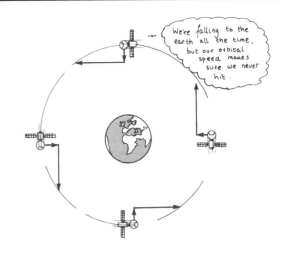

20 Draw a free-body diagram for a spaceman in Earth orbit. Does he need a force to maintain his orbit speed? Explain why he can 'float' inside his craft even though his body has weight.

Inertia

Getting things to move

Imagine you are trying to set the three objects above into motion. The one with the most mass would be the most difficult to get started. This is not because of friction. Even floating in space, the car would need nearly as much effort to start it moving. All bodies resist being set into motion, especially ones with large mass. Once moving, these massive bodies are also difficult to stop. This property of mass is called **inertia**.

As an example, imagine a fat man and a thin man using identical rocket engines to get moving. The fat man gets less acceleration and takes longer to gain speed than the thin man. It is more difficult to get him moving because of his great mass. The fat man has more inertia than the thin man.

Momentum

When we stop a moving ball we feel a blow. If the ball is large and moving fast, the blow is bigger. The ball has **momentum**. (The quantity of momentum is mass × speed.) The more momentum an object has, the bigger the blow it delivers when it is stopped. When we kick a ball, the blow gets the ball moving and gives it momentum. The bigger the blow, the greater is the momentum gained by the ball.

21 Write down these moving objects in order, with the one with greatest momentum first and so on: a lorry travelling at 80 km/h; a boy cycling at 30 km/h; an oil-tanker sailing at 15 km/h; a golf ball in flight at 100 km/h.

22 For each example name:
(a) an object that loses momentum;
(b) an object that gains momentum as a result of a blow.

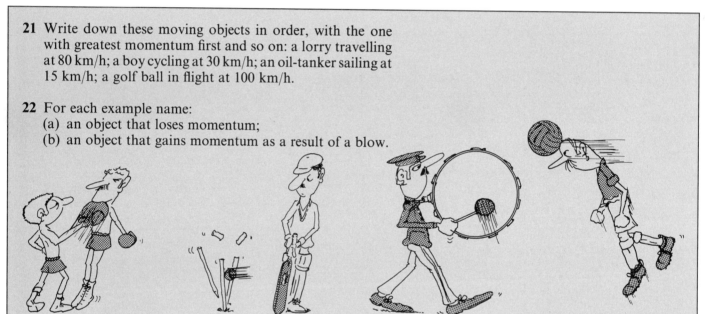

More about momentum

When you flick a paper pellet across a table, you strike it with a large force for a short time. The flick is a blow or impulse that is measured by multiplying the force × the time it acts. The impulse gives the pellet momentum. The impulse and the momentum it produces are both vectors and are found to be equal. So

$$\text{impulse} = \text{momentum gained}$$

$$\text{force} \times \text{time} = \text{momentum gained}$$

$$\text{force} = \frac{\text{momentum gained}}{\text{time it acts}}$$

And in general, force = the rate of change of momentum

This is another form of Newton's second law of motion (see p. 41).

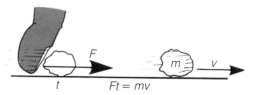

Impulse = momentum gained

Force and Motion check list

After studying the section on Force and Motion you should be able to:

- use a ticker-timer to measure time and speed
- describe a motion from dots on ticker-tape
- make a speed/time graph from ticker-tape
- plot distance/time graphs from readings
- calculate speed from a distance/time graph
- interpret distance/time graphs
- know how to use the equation
  average speed = distance covered/time taken.
- plot speed/time graphs
- calculate acceleration and distance travelled from speed/time graphs
- know how to interpret speed/time graphs
- know the difference between velocity and speed
- know what is meant by positive and negative acceleration
- know how to make a ticker-tape graph of the acceleration of an object
- know how to calculate acceleration
- know three equations of uniform acceleration and how to use them
- understand the way that force affects motion
- know Newton's first law of motion
- know how to show that acceleration gets larger if the accelerating force is increased
- know how to find out if force and acceleration are proportional
- know the equation that connects force, mass and acceleration
- know how to use this equation
- know how the newton is defined
- know Newton's second law of motion
- know that the acceleration of gravity is the same for all objects at the same place
- know how to show this without air resistance
- know how to measure the acceleration of gravity
- know the meaning of gravitational field strength
- understand the meaning of 'weightlessness' on Earth and in orbit
- understand the idea of inertia
- understand the idea of momentum

Revision quiz

Use these questions to help you revise Force and Motion.

- How can you measure speed from a piece of printed ticker-tape?
  ... measure the distance covered by the dots and calculate the time from the number of dots (50=1 s).

- How can you make a speed/time graph from ticker-tape?
  ... cut the tape into 5-dot lengths and stick them side by side.

- How can you tell if the motion is accelerating?
  ... the dots get further apart and the 5-dot pieces get longer and longer.

- How can you measure speed from a distance/time graph?
  ... speed is the gradient of the line.

- What is the equation for average speed?
  ... average speed=distance travelled/time taken.

- Rearrange this to give an equation for distance travelled.
  ... distance=speed×time.

- Rearrange again to give an equation for time.
  ... time=distance/speed.

- How can you calculate acceleration from a speed/time graph?
  ... acceleration is the gradient of the line.

- How can you calculate the total distance travelled from a speed/time graph?
  ... distance travelled=the total area under the graph.

- What is the difference between speed and velocity?
  ... velocity is speed in a named direction.

- What is meant by negative acceleration?
  ... when the acceleration acts in the opposite direction to the velocity vector.

- Give an equation for calculating acceleration.
  ... acceleration=change in velocity/time taken.

- Quote three equations of uniformly accelerated motion.
  ... (1) $v=u+at$ (2) $s=ut+\frac{1}{2}at^2$ (3) $v^2=u^2+2as$

- What is Newton's first law of motion?
  ... if the forces on a body balance then it will not move, or if it is already moving it will continue to move at a steady speed.

- What is Newton's second law of motion?
  ... the acceleration of a body is proportional to the net force that acts on it provided its mass does not change.

- Write Newton's second law as an equation.
  ... force=mass×acceleration.

- Define the newton.
  ... one newton is the force required to give a mass of 1 kilogram an acceleration of 1 m/s^2.

- Is the acceleration of gravity the same for light and heavy objects?
  ... yes, however lighter objects will fall more slowly due to air resistance.

- What is gravitational field strength?
  ... it is the force per kilogram that acts on objects in a gravitational field.

- What is meant by 'weightlessness' on or near the Earth?
  ... it is when people do not have to support their weight because they are free-falling to the ground.

- What is meant by the inertia of an object?
  ... inertia is the property of a mass that makes it resist being moved or being stopped if it is already moving.

- What is meant by the momentum of a moving object?
  ... the momentum of an object is its mass×velocity and is a measure of how big a blow is needed to stop it moving.

Examination questions

1. A man runs a race against a dog. Fig. 1 is a graph showing how they moved during the race.
   (a) What was the distance for the race?
   (b) After how many seconds did the dog overtake the man?
   (c) How far from the start did the dog overtake the man?
   (d) What was the dog's time for the race?
   (e) Use the equation $v = s/t$ to calculate the average speed of the man.
   (f) After 8 seconds is the speed of the man increasing, decreasing or staying the same?
   (g) What is the speed of the dog after 18 seconds?

   (*NEA*)

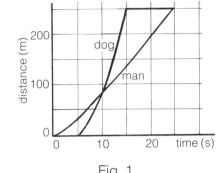

Fig. 1

2. The sketch graph in Fig. 2 represents a journey in a lift in a department store.
   (a) Briefly describe the motion represented by
       (i) OA, (ii) AB, (iii) BC.
   (b) Use the graph to calculate
       (i) the initial acceleration of the lift,
       (ii) the total distance travelled by the lift,
       (iii) the average speed of the lift for the whole journey.

   (*NEA*)

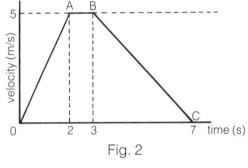

Fig. 2

3. (a) A car engine is leaking oil. The oil drops hit the ground at regular time intervals, one every 2.0 seconds.
   The diagram below shows the pattern of the drops that the car leaves on part of its journey.
   (i) What can you say about the speed of the car before it reaches the signs?
   (ii) Calculate the distance between the drops on the road before it reaches the signs if the car is travelling at 10 m/s.
   (iii) After the car passes the signs, what happens to the gaps between the drops of oil?
   What does this tell you about the motion of the car?
   (iv) Further down the road it is found that the distance between the drops on the road has become 30 m. What is the speed of the car at this point?

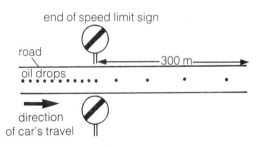

Fig. 3

   (b) A front-wheel drive car is travelling at constant velocity. The forces acting on the car are shown in the diagram below. F is the push of the air on the car.
   (i) Name the 400 N force to the right.
   (ii) What is the value of the force F to the left?
   (iii) Taking the weight of 1 kg to be 10 N, calculate the mass of the car.
   (iv) The force to the right is now increased. Describe and explain what effect this has on the speed of the car.

   (*LEAG*)

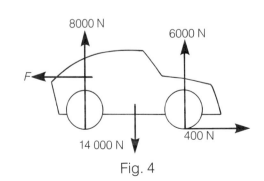

Fig. 4

4 Slow motion photography shows that a jumping flea pushes against the ground for about 0.001 s, during which time its body accelerates upwards to a maximum speed of 0.8 m/s.
  (a) Calculate the average upward acceleration of the flea's body during this period.
  (b) If the flea then moves upwards with a constant downward acceleration of 12 m/s² find
    (i) how long it will take, after leaving contact with the ground at a speed of 0.8 m/s, to reach the top of its jump,
    (ii) how high it will jump after leaving contact with the ground.
  (c) Why is the acceleration of the flea after leaving the ground not equal to g?

(*LEAG*)

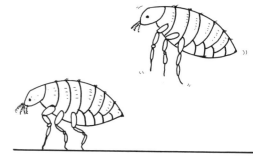

Fig. 5

5 This question is about SPEED and ACCELERATION.
A cycle track is 500 metres long. A cyclist completes 10 laps (that is, rides completely round the track 10 times).
  (a) How many kilometres has the cyclist travelled?
  (b) On average it took the cyclist 50 seconds to complete one lap (that is, to ride round just once).
    (i) What was the average speed of the cyclist?
    (ii) How long in minutes and seconds did it take the cyclist to complete the 10 laps?
  (c) Near the end of the run the cyclist put on a spurt. During this spurt it took the cyclist 2 seconds to increase speed from 8 m/s to 12 m/s.
What was the cyclist's acceleration during this spurt?

(*SEG*)

6 This question is about FORCE and ACCELERATION.
The driver of a car moving at 20 m/s along a straight level road applies the brakes. The car decelerates at a steady rate of 5 m/s².
  (a) How long does it take the car to stop?
  (b) What kind of force slows the car down?
  (c) Where is this force applied?
  (d) The mass of the car is 600 kg.
What is the size of the force slowing the car down?

(*SEG*)

7 A pendulum was made from light twine and a lead ball. The length of the pendulum was varied and the number of complete swings it made in one minute recorded. The results are shown in this table.
  (a) On graph paper, plot a graph of swings per minute vertically against length of pendulum horizontally.
  (b) Write down a simple conclusion about the number of complete swings per minute and the length of the pendulum.
  (c) By drawing on the graph, find the length of a pendulum which makes one complete swing in one second.

(*NEA*)

Length (cm)	Number of swings per minute
10	95
20	67
30	55
45	45
50	42
60	39

6 Pressure

You cannot push your thumb into a table but with the same force you can push a drawing pin into the wood. To explain how the same force can have such different effects we must consider the area on which the force presses. When the area is small, the force makes a large dent in soft substances. The force produces a large pressure if the area is small. Pressure is the force that acts on 1 m² and is calculated by dividing force by area.

$$\text{pressure} = \frac{\text{force}}{\text{area}}$$

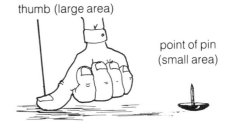

1 Copy the table and for each object say whether it has a 'large' or 'small' contact area and whether the pressure under it is 'large' or 'small' when it is being used.

Object	Area	Pressure
Nail		
Ice skates		
Hippo's feet		
Knife edge		
Caterpillar tracks		

EXPERIMENT. Measuring your pressure on the ground

Find your weight (in newtons) from a balance or scales. Draw round your feet on squared paper. Find the area you stand on by counting the squares inside the outline of your shoes. Calculate your pressure on the floor (weight ÷ area).

Example

A boy weighs 500 N and the area of both his shoes came to 250 cm². His pressure on the floor = 500 ÷ 250 = 2 N/cm².

The area should be in square metres. As there are 100 × 100 centimetre squares in a 1 metre square, the above pressure is 2 × 100 × 100 N/m² = 20 000 N/m² = 20 kN/m².

The unit N/m² is called the pascal (Pa).

$$\text{pascal} = \text{newton/square metre}$$

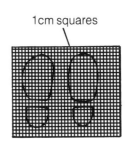

Measuring your pressure

2 Which of these people would probably cause the greatest pressure on the ground:
   (a) a fat man with big feet,
   (b) a fat man with small feet,
   (c) a thin girl with big feet,
   (d) a thin girl with small feet.

3 Copy and complete this table.

Force (kN)	Area (m^2)	Pressure (kPa)
200	2	
50		10
	4	8

4 A boy pushes his thumb onto a table with a force of 12 N. His thumb has a contact area with the table of 4 cm^2. He then pushes with the same force on a drawing pin that has a contact area of 1 mm^2 ($=\frac{1}{100}$ cm^2). Calculate the pressure exerted on the surface by (a) his thumb, (b) the drawing pin.

Pressure in liquids

The water in a home aquarium is surprisingly heavy; it probably weighs more than you do. Its weight presses down and exerts a pressure on the bottom of the tank. You could calculate this pressure by dividing the weight of the water by the area of the bottom of the tank.

The pressure of the water also acts sideways on the walls of the tank. You would see this if you made holes in the walls. The pressure would force jets of water to squirt out sideways.

The pressure in the water gets greater as you go deeper. This is because there is more water above you to press down. The jets of water show this by squirting out further from the lower holes.

Pressure also acts all over the surface of objects that are immersed in the water.

5 The figures show the calculation of mass, pressure, weight and volume of water in the aquarium. Copy the table and write these quantities in the correct spaces. Also put units against each quantity.

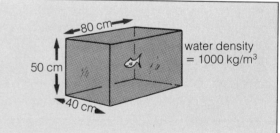

Quantity	Calculation	Unit
	$0.8 \times 0.4 \times 0.5 = 0.16$	
	$1000 \times 0.16 = 160$	
	$160 \times 10 = 1600$	
	$1600/(0.8 \times 0.4) = 5$	

An equation for the pressure of a liquid

The pressure deep in a liquid = force/area, where the force is the weight of liquid that lies above the chosen area. If this area is 1 m² and the depth of the liquid is h, then the force is the weight of liquid in a column of height h.

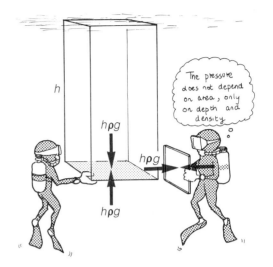

> The pressure does not depend on area, only on depth and density.

The volume of this column = $h \times 1$.
The mass of liquid it contains = $h \times 1 \times \rho$ (where ρ = density of the liquid, see p. 299).
The weight of this liquid = $h \times \rho \times g$.
This is the force on 1 m² area so . . .

> The pressure at this depth = $h\rho g$

Standard atmospheric pressure

'One standard atmosphere' is the pressure that can support a column of mercury 760 mm high. For mercury $\rho = 13\,590$ kg/m³ and $g = 9.81$ N/kg. (See also p. 45.)

So the pressure of this mercury column = $0.76 \times 13\,590 \times 9.81 = 101\,321$ N/m². Standard atmospheric pressure, 760 mmHg or 1 atm, is therefore a pressure of 101 321 N/m².

For these questions take the density of water as 1000 kg/m³ and $g = 10$ N/kg.

6 What is the pressure due to water, at the bottom of a 3 m deep swimming pool?

7 How deep must you go for water to exert a pressure of 1 atm (101 321 Pa).

8 What is the pressure (in atmospheres) on the wreck of the Titanic that lies at a depth of 3.8 km in sea water. (Density = 1100 kg/m³.)

9 This party trick uses air pressure on a card to keep water in a glass. If the card weighs 10 g, calculate the force that holds it onto the glass.

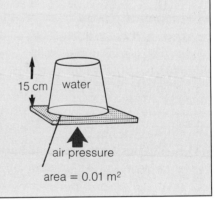

Pressure through liquids

Fill a bicycle pump with water, put your finger over the end and try to compress the water. You will find that water cannot be squashed into a smaller volume and that the force on the handle passes through the water to your finger.

EXPERIMENT. Liquids transmit pressure

Connect two syringes together as shown. If you push in both of the plungers you feel that the water is 'rigid' and not squashy. The pressure from one piston goes through the water and acts on the piston of the other syringe. You will also notice that one piston needs more force to hold it than the other.

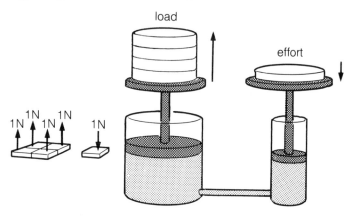

Transmitting pressure through a liquid

> **10** Explain which piston has to be held with most force and why.

A simple hydraulic machine

Although the pressure is the same on both pistons, more force acts on the large piston than on the small one. A hydraulic press uses this idea to increase force. Two cylinders of different sizes are connected by a pipe full of liquid. A load is put on the large piston and the effort is pressed on the small piston. You will find that a small weight can lift a large weight with this machine.

A hydraulic press

Why does it work?

Pressure produced by the effort passes through the liquid to the load's piston. This piston has a large area and so the pressure presses on a large area. This produces a large force. If the load piston has four times the area of the effort piston, the force on it will be four times as much.

> **11** Copy this table and write large, small or same in the gaps.
>
> | | Effort | Load |
> |---|---|---|
> | Size of force | | |
> | Area of piston | | |
> | Pressure on piston | | |
>
> **12** If the effort's piston moves down, the load's piston moves up but not as far. Why does the large piston move less than the small piston?

13 This picture shows how all the brakes of a car are connected by pipes to a 'master' cylinder. The pipes and cylinders are full of light oil. When you press the foot brake, pressure passes through the oil to the brake cylinders. Pistons inside move, putting on all the brakes of the car together.
   (a) Would the front brakes still work if the pipe got blocked at X?
   (b) What difference would it make to the movement of a piston if one of the cylinders had air in it instead of oil (an air-lock)?
   (c) The brake cylinders are larger than the master cylinder. Why is this an advantage to the driver?

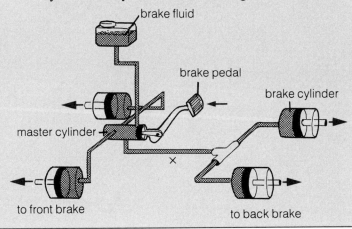

Car brakes – drum type

Car drum brakes have two curved brake shoes placed just inside the rotating wheel drum. These shoes can be moved outwards at the bottom by the piston in the brake cylinder. When the foot pedal is pushed, pressure passes through the liquid and moves that piston so that the brake shoes move onto the wheel drum. Friction then stops the wheel from going round. Note that water is not used as the liquid in hydraulic machines because it evaporates rather easily and can contain dissolved air. This air may form air bubbles or dangerous air-locks in the pipes.

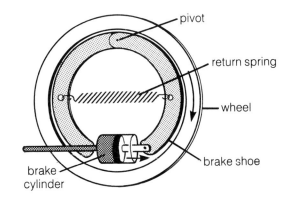

Car brakes – disc type

To make a more effective brake, a thick disc of metal is bolted to the wheel. As the wheel goes round, this disc rotates between two brake pads. When the brake pedal is pushed, these pads are forced onto the disc from each side. Friction then acts on the disc in the opposite direction to its motion and stops the wheel. The pads can be made to pinch the disc with great force because they are connected to the pedal by a hydraulic system. The large surface area of the disc helps to keep the brakes cool by removing heat from the friction pads.

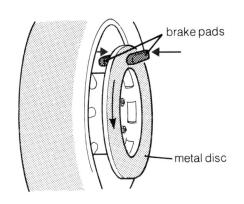

Other hydraulic machines

This (man-powered) hydraulic jack has two valves. It is often used in garages to lift cars and lorries. The valves make it possible to repeat the movement of the effort handle and raise the load in stages through large distances.

When the effort handle moves down it forces liquid into the load cylinder. When the effort piston moves up, liquid refills the effort cylinder. The movement can then be repeated.

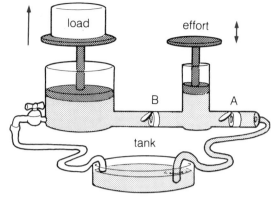

A hydraulic jack

14 Copy and insert 'open' or 'closed' in the boxes.

	Valve A	Valve B
Effort piston moving down		
Effort piston moving up		

15 What forces liquid to move up from the tank to the effort cylinder?

16 What use has the tap? Why is it a good idea to connect a pipe from the tap to the tank?

A motor-powered hydraulic machine

Farmers, pilots, bulldozer drivers, building workers and many others use a type of hydraulic machine that is driven by a pump. The pump forces liquid round a circuit. All the operator has to do is to use control valves to direct the path of the liquid.

A pump drives oil round a circuit ABCD. A control valve can then send liquid into either end of a working cylinder. The operator can thus move the piston in this cylinder to the right or left with immediate control and great force.

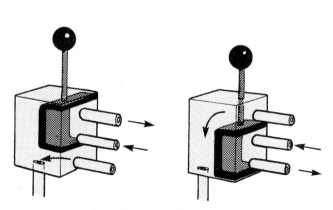

How the control valve works

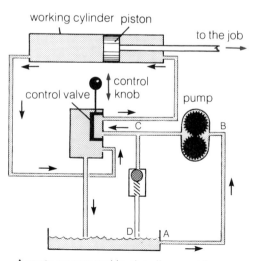

A motor powered hydraulic machine

17 Draw the simple picture of this machine with the control valve covering the bottom two pipes as shown. Draw arrows to show the flow of liquid into the cylinder. Which way would the piston move?

18 What two differences would it make if a larger working cylinder were used? (The pump is not changed.)

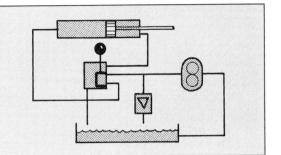

Manometers

A manometer is a glass U-tube containing liquid, that is used to measure gas pressure. The gas forces liquid down one arm and up the other until the pressure of the liquid balances the pressure of the gas. The gas pressure is measured by the distance between the liquid levels in centimetres or metres.

Mercury is a liquid that can be used for high pressures and water or oil for low pressures.

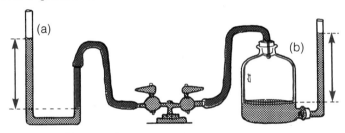

Manometers measuring gas pressure

EXPERIMENT. Using manometers

(1) Use a small water manometer to measure the pressure of the 'gas'.

(2) Make a manometer with arms of different width. You could join a wide tube to a narrow one or connect a bottle to a tube. Use this manometer to measure the same 'gas' pressure.

You should get the same result because the pressure of the liquid in these tubes does not depend on their area.

(3) Blow up a plastic bag and connect it to a manometer. Put a 1 kg weight on a board on the bag and measure the pressure of the air. Notice how the pressure of the air depends on the size of the board.

'Blowing up' a heavy weight

(4) Use a large water manometer to measure your lung pressure. (Do not blow too hard.)

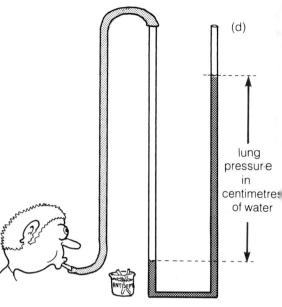

A manometer for measuring lung pressure

19 How would you use manometer (a) to tell if there was a gas leak in the rubber tube connecting it to the tap?

20 What will happen when the gas tap is turned on if the gas pressure is 5 cm of water?

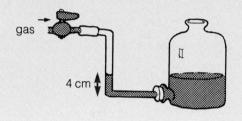

The pressure of the atmosphere

EXPERIMENT. Does the air around exert a pressure on us?

Take an old metal can and fit it with a cork and tube. Connect the tube to a vacuum pump and remove the air from inside the can. You will see the can crumple under the great pressure of the outside air. Normally the air pressure inside balances the air pressure outside but by removing the air inside we see what the outside pressure can do.

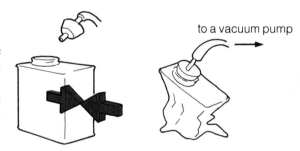

The crushing effect of air pressure

EXPERIMENT. How great is the pressure of the air around us?

Take a large mercury manometer and remove the air from one side by connecting it to a vacuum pump. You will see that the air on the other side had enough pressure to make the mercury rise about 76 cm ($=100$ kN/m^2).

This is a considerable pressure (the weight of about 200 people on 1 square metre or the weight of 1 kg on 1 square cm). The pressure of the air around is called atmospheric pressure.

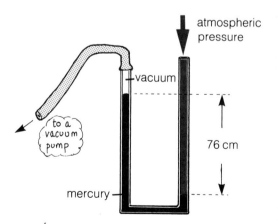

Measuring atmospheric pressure

EXPERIMENT. More evidence of the atmosphere's great pressure

You may be able to do this old experiment in the lab. The apparatus consists of two strong, hollow hemispheres of metal that fit together with an air-tight seal. There is a tube that allows you to remove the air from the inside chamber and a tap to stop the air from returning.

If you make a good vacuum inside you will find it impossible to pull the two halves apart.

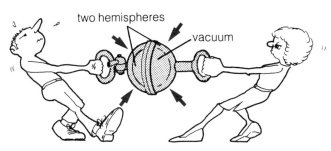

(there are no screws or glue holding the two halves together)

Pulling against atmospheric pressure

It cannot be the vacuum that is 'sucking' the two halves together – a vacuum means nothing is there. It must be the pressure of the atmosphere outside that presses the two halves so tightly together.

21 How does this experiment show that atmospheric pressure acts upwards and sideways as well as downwards?

22 Explain why liquid fills a syringe when the plunger is pulled out and the end is under the liquid.

23 When you suck up liquid with a straw, it is atmospheric pressure that pushes liquid up the straw. Why is it impossible to suck liquids from the bottle on the right?

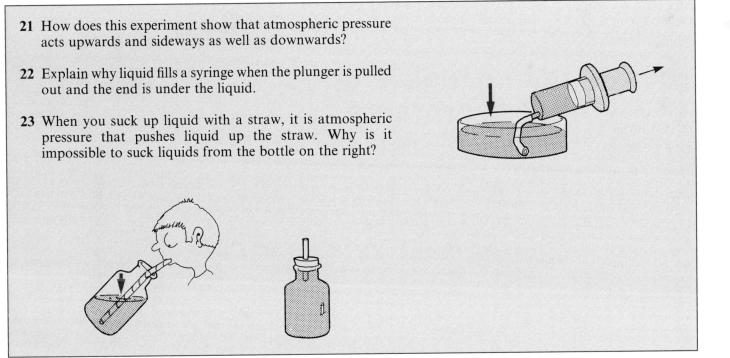

Aneroid barometer

A barometer measures atmospheric pressure. As atmospheric pressure changes with the weather, barometers are used in weather forecasting. High pressure usually means fine weather and low pressure bad weather.

An aneroid barometer uses a thin metal can that has had air taken out of it. A strong spring stops the can from being crushed by atmospheric pressure. If the atmospheric pressure rises, the lid of the can is pushed in slightly. If the pressure drops, the spring pulls the lid up a little. The small movement is magnified by a pointer or gears.

This sort of barometer can be fitted into an aeroplane and flown to high altitudes. Since atmospheric pressure drops as you go up, the barometer reading can be used to show the height of the plane. It is then called an altimeter.

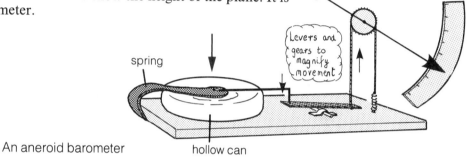

An aneroid barometer

24 Draw this diagram of an aneroid barometer. If the atmospheric pressure rises, pushing the lid of the can down, which way will the end of the pointer move? Write 'high pressure' and 'low pressure' on the scale.

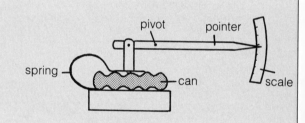

The Bourdon gauge – for measuring gas pressure

The gauge contains a hollow metal tube. When the pressure rises in this tube, the tube tries to straighten and the pointer rotates. This is rather like the way a rolled up paper tube, containing a spring, straightens when you blow into it.

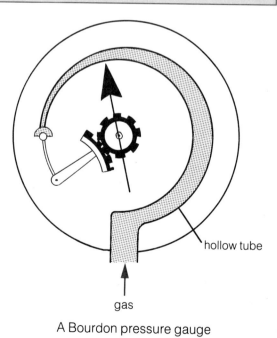

A Bourdon pressure gauge

Air pumps

We find it very useful to compress air and so increase its pressure.

A cycle pump and valve can force air into a tyre or football and give them bounce. Take a pump apart and look at its parts.

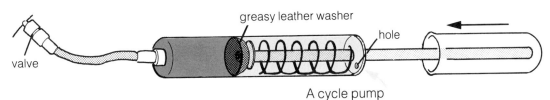

A cycle pump

The key working part of a cycle pump is a strong, flexible washer made of leather or plastic. As the pump handle is pushed in, this washer presses against the walls and forms an air-tight seal. The air in the barrel is then forced out through a valve. When the handle is pulled out, the washer collapses and air moves past it, refilling the barrel. In this way more and more air can be forced through the valve.

25 Put these sentences, describing what happens when the pump handle is moved IN and then OUT, in order.
   (a) Air in A is compressed and forced out.
   (b) Air squeezes past the edge of the washer from B to A.
   (c) Outside air goes through the hole into B.
   (d) A partial vacuum forms in A.
   (e) The washer is pressed hard against the barrel of the pump.

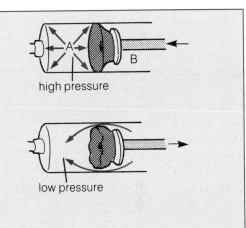

26 Why is the valve necessary? What would happen if it was not there?

27 How can you alter this pump to make it suck instead of blow?

28 How would you change the shape of the barrel to make a pump that can:
   (a) produce a higher pressure,
   (b) pump up tyres with fewer strokes?

29 Describe how you could measure the volume of air pushed out by a pump in one stroke.

Compressed air

Try squeezing the air in a bicycle pump. Notice how 'springy' it is. Note how increasing the pressure forces the air into a smaller volume.

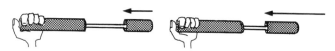

Air can be compressed

EXPERIMENT. How does the volume of some trapped air change as the pressure on it is increased?

Use this or similar apparatus to get readings of the volume of some air and the pressure on it.

The air is trapped in a tube by oil. The tube is marked so that the volume of the air can be measured. A cycle or foot pump is used to compress the air and the pressure is read on a Bourdon gauge.

Use the numbers you get to look for a connection between the pressure of the air and its volume.

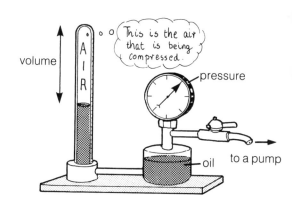

Boyle's law

Robert Boyle put forward a law connecting pressure and volume. The law says that:

the volume of a fixed mass of gas is inversely proportional to its pressure, provided its temperature does not change.

The readings in this table make the meaning of this clearer. If the pressure is doubled the volume is halved; if the pressure is trebled then the volume becomes one third of what it was and so on. This is called inverse proportion.

Notice that if pressure and volume readings are multiplied together you get the same answer (1 in this case). This is a way of testing for inverse proportion.

Pressure	1	2	3	4	5
Volume	1	$\frac{1}{2}$	$\frac{1}{3}$	$\frac{1}{4}$	$\frac{1}{5}$
Pressure × volume					

So Boyle's law can be stated as a word equation:

$$\text{pressure} \times \text{volume} = \text{a constant}$$

Test your readings for inverse proportion. Does the air in your experiment follow Boyle's law?

---

30 Plot graphs of pressure against volume for
  (a) the readings in the table above,
  (b) your own readings.

31 (a) Do the readings in this table follow Boyle's law?
   (b) What would the volumes be for pressures of 110 kPa, 130 kPa and 150 kPa.

Pressure, P (kPa)	100	120	140	160
Volume, V (ml)	50	41.7	35.7	31.3
$P \times V$				

Hot air

If the air in a tin can is heated and it cannot expand, there will be an explosion. Heating air is another way of increasing its pressure.

EXPERIMENT. Heating and cooling air

Take readings of the pressure and temperature of air as it is heated and cooled. Seal the air in a strong flask with a bung that is fitted with a pressure gauge. Heat the flask in a water-bath and take readings of the pressure of the air and its temperature. Take further readings when the air is cooled in an ice-bath.

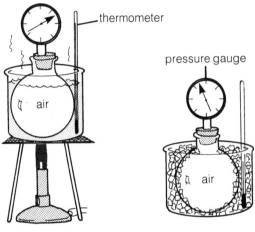

Looking for a pattern

Here are readings from such an experiment. Use them (and your own results) to look for a connection between pressure and temperature.

Pressure (kN/m^3)	128	121	114	107	100	93	89
Temperature (°C)	100	80	60	40	20	0	−10

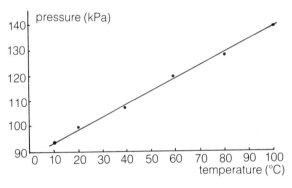

Plot two graphs of the readings:

(a) One on axes like these (on the right):

This should show that the pressure rises steadily with temperature.

(b) Another on axes like these:

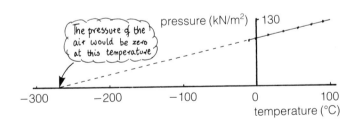

These axes allow you to continue the line backwards and find out how cold you must make the gas to lower its pressure to zero. Use your readings to get a value for this temperature.

The temperature (in theory) is −273 °C (273 °C below zero). This is as cold as anything can be and is called the **absolute zero of temperature**.

The Kelvin temperature scale

Sometimes temperatures are measured on a scale that starts at absolute zero. This is called the **Kelvin scale**. The ice point (0 °C) would be 273 K on this scale; the steam point (100 °C) would be 373 K. Temperatures on the Celsius scale can be converted to the Kelvin scale by adding 273.

The Pressure law

This law says that:

the pressure of a fixed mass of gas is proportional to its Kelvin temperature, provided its volume doesn't change.

The simplified readings in this table make this clear. The pressure doubles if the Kelvin temperature doubles and so on. Pressure and Kelvin temperature are directly proportional. One test for direct proportion is to divide temperature by pressure. The answer will be the same each time (2 in this case).

Temperature (°C)	−173	73	37	127
Temperature (K)	100	200	300	400
Pressure (kPa)	50	100	150	200
Temperature (K) / Pressure (kPa)	2	2	2	2

32 Copy this table of readings and
   (a) calculate the Kelvin value of each temperature.
   (b) calculate Kelvin temperature/pressure in each case.
   Do the readings follow the Pressure law?

Pressure (kPa)	128	121	114	107	100	93	89
Temperature (°C)	100	80	60	40	20	0	−10
Temperature (K)							
Temperature (K) / Pressure (kPa)							

33 Describe a simple experiment to get a value for the absolute zero of temperature.

34 One way of getting dents out of ping-pong balls is to put them in boiling water.
   Explain why the dents get pushed out.

Pressure check list

After studying the section on Pressure you should:

- know how to calculate the pressure of a force on a surface
- know that pressure is measured in pascals
- know how pressure acts in a liquid
- know how to calculate pressure at a depth in a fluid
- know that liquids transmit pressure
- understand hydraulic machines and car braking systems
- know how to make and use manometers to measure gas pressure
- know that the atmosphere exerts a large pressure
- know how to measure and calculate atmospheric pressure with a mercury manometer
- know how aneroid barometers and Bourdon gauges measure pressure
- know how a bike pump works
- know how to investigate the pressure and volume of air and test if it follows Boyle's law
- know how to investigate the pressure and temperature of air and check if it follows the Pressure law
- understand the idea of absolute zero of temperature and the Kelvin temperature scale

Revision quiz

Use these questions to help you revise the section on Pressure.

- What is the formula for calculating pressure? — ...pressure=force/area.
- What are the units of pressure? — ...newton/square metre which is the same as pascal.
- How does pressure act in a still liquid? — ...it acts in all directions, it is the same along a horizontal level, it gets greater with depth and the density of the liquid.
- What is the formula for the pressure at depth h in a liquid of density ρ? — ...pressure=$h\rho g$.
- What property of liquids makes them useful in hydraulic machines? — ...they cannot be compressed but can pass on pressure from one piston to another.
- Why does a hydraulic machine increase the force on a load? — ...because the load is placed on a larger piston than the effort.
- What is a manometer? — ...it is a transparent U-tube, partly filled with liquid, that is used to measure gas pressures.
- How is pressure measured on a manometer? — ...pressure is measured as the difference in liquid levels.
- Why is a mercury manometer needed to measure atmospheric pressure? — ...because atmospheric pressure is so great. It causes a difference in levels of about 76 cm of mercury.
- Name two other instruments that measure pressure — ...the aneroid barometer and the Bourdon gauge.
- Name two ways of increasing the pressure of air. — ...compressing it with a pump and heating it without letting it expand.
- What is Boyle's law? — ...the volume of a fixed mass of gas is inversely proportional to its pressure provided its temperature does not change.
- Give Boyle's law as an equation. — ...pressure×volume=a constant number.
- What is the Pressure law? — ...the pressure of a fixed mass of gas is proportional to its Kelvin temperature provided its volume does not change.
- What is meant by the absolute zero of temperature? — ...it is the temperature at which the pressure of a gas (in theory) becomes zero.
- What is the absolute zero of temperature? — ...−273 °C
- What is the Kelvin temperature scale? — ...it is a scale that measures temperatures from the absolute zero
- How do you convert from the Celsius to the Kelvin scale? — ...add 273.

Examination questions

1 A rectangular block measures 8 cm by 5 cm by 4 cm, and has a mass of 1.25 kg.
  (a) (i) If the gravitational field strength is 10 N/kg, what is the weight of the block?
      (ii) What is the area of the smallest face of the block?
      (iii) What pressure (in N/cm^2) will the block exert when it is resting on a table on its smallest face?
      (iv) What is the least pressure the block could exert on the table?
  (b) (i) What is the volume of the block?
      (ii) Calculate the density of the material from which the block is made.
                                                                                    (NEA)

2 (a) Describe how you would show experimentally the relationship between the volume and pressure of a fixed mass of gas at constant temperature.
  Your answer should include
    (i) a diagram;
    (ii) a statement of the readings made;
    (iii) one precaution taken to ensure a reliable result.
  (b) The following readings for the pressure and temperature of a fixed mass of gas were obtained in a different experiment.

Pressure (kPa)	3.80	4.08	4.36	4.64	4.92	5.20
Temperature (°C)	0	20	40	60	80	100

  (i) Plot a graph of pressure (*y*-axis) against temperature (*x*-axis).
  (ii) From the graph read and state the temperature when the pressure is 4.50 kPa.
  (iii) Find the pressure when the temperature is 150 K.
  (iv) Comment on the reliability of the values obtained in parts (ii) and (iii).
                                                                                    (LEAG)

3 This is a question about hydraulic jacks.
  (a) Fig. 1 represents a simple form of hydraulic jack. (The pistons may be considered as weightless and frictionless.) A force of 10 N is applied at A.
    (i) Explain why the downward force at A causes the load to rise.
       Calculate the pressure, due to the 10 N force at A, exerted
    (ii) by the small piston on the liquid at B,
    (iii) by the liquid on the wall at C, and
    (iv) by the liquid on the large piston.
    (v) Calculate the maximum load that can be raised.

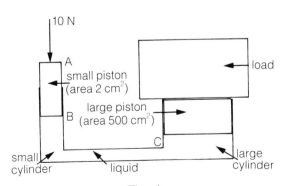

Fig. 1

(b) Fig. 2 shows a practical form of hydraulic car jack. When the lever, and so the pump piston, is moved to the right the large piston moves upwards.

Explain why
  (i) when the lever is moved to the left the large piston does *not* move downwards,
  (ii) when the release valve is opened by partially unscrewing it, a jacked up car is lowered to the ground.

(*LEAG*)

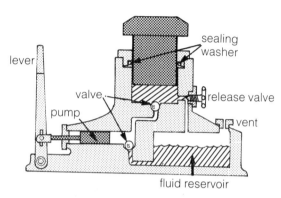

Fig. 2

Section 2
ENERGY

7 Energy

Energy can be stored in a number of forms

The pictures on this page show things that have a supply of stored energy.

The cyclist has energy stored in his muscles. The chemicals in his muscles supply the energy and change when the muscles are used. This form of stored energy is called **chemical energy**.

The motorbike engine uses energy stored in the petrol. The petrol has chemical energy because its chemicals change when it is burnt.

A jumping cracker also has stored chemical energy, that can be released and used to make the cracker jump about.

A skier or trolley at the top of a hill has a different form of stored energy. This energy can be used to move them downhill at speed. Things that are high up have what is called **gravitational energy**. This is energy that is there ready to be used because of the pull of gravity.

Energy can also be stored in stretched elastic or springs. The stretched elastic of a catapult has stored energy which can be used to move a stone. This is called **strain energy**.

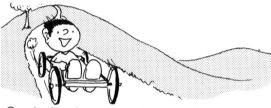
Chemical energy getting bikes moving

Gravitational energy getting a trolley moving

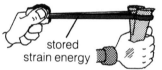
Strain energy getting a stone moving

You will notice that all these forms of stored energy can be used to get things moving. A way of recognizing energy is by its ability to get things moving.

Potential energy

Chemical, gravitational and strain energies can be stored by things; the energy is held by them, waiting to be used. Energy that can be held in readiness is called **potential energy**. Chemical, gravitational and strain energies are all potential energies.

Forms of potential energy

1 Copy this table and write in the form of stored energy that each object has.

Object	Form of stored energy
Stretched chest expander	
Cyclist	
Hammer on a table	
Firework	
Motorbike engine	
Trolley on a hill	
Stretched catapult	

2 Give an example of something that has
   (a) gravitational potential energy due to its high position;
   (b) strain potential energy because it is stretched;
   (c) chemical potential energy.
   In each case explain how the energy can be used to produce movement.

Kinetic energy

Objects that are moving can make other things move. The energy of a moving object is called **kinetic energy** (kinetic means 'moving').

We have seen how the three forms of potential energy can make things move. That is, they change into kinetic energy. The following experiment shows energy changing between kinetic and potential forms.

EXPERIMENT. An energy converter

Make a pendulum by tying a 10 g mass onto a thread and clamping the top end of the thread between two wooden blocks. Raise the mass by pulling it to one side (this supplies it with a little gravitational energy) and then let go.

The bob will swing down, changing its gravitational energy into kinetic energy. Then, as it swings up, the kinetic energy changes back into gravitational energy. The energy continues to change form and the pendulum vibrates from side to side.

Pendulum power

Pull the 10 g mass to one side so that it rises 1 cm. This will give it 1 mJ of gravitational energy ($=mgh$, see p. 85). Release the mass and find out how long this small amount of energy will keep the mass in motion.

The moving mass loses energy as it stirs up the air. Work out how much energy is lost to the air each second by dividing 1 mJ by the time taken for the pendulum to stop. This is the power of the air's resistance.

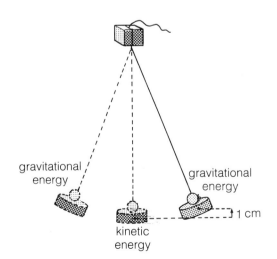

> 3 Explain why the power calculated above is an average value. Suggest another way that the pendulum can lose its energy.

Other forms of energy

Heat, light and sound are forms of energy that move from one place to another. **Heat** carries energy from things that are hot to things that are not so hot; **light** carries energy away from lamps or the sun; **sound** carries energy away from the vibrations of a noisy object.

Electrical energy is a form that is easy to move around. We use it to bring energy into our homes and work places, where we change it into the forms we want.

Heat, light and sound are all forms of energy

> 4 Write a list of the forms of energy mentioned on pp. 72–74. There are eight.
>
> 5 Give an example of an effect caused by each form of energy.
>
> 6 What forms of energy are needed to:
> (a) clean your teeth;
> (b) make a tape recorder work;
> (c) toast a slice of bread;
> (d) make water run out of a tap?
>
> 7 What form(s) of energy do the following have: a live match, a gallon of petrol, wind, waves, food, a chimney pot on top of a house, water at the top of a waterfall, an air gun?

More energy changes

Energy can be changed freely from one form to another. Energy that is stored can be released and changed into forms that do useful jobs for us. Energy however cannot be converted completely into just one other form. Some is always changed into less useful forms such as heat.

Here are some examples of energy conversions.

1. Throwing a dart

The lad on the next page uses some of his stored energy to move a dart. His muscles use some of their chemical energy to give the dart kinetic energy. Not all of the chemical energy turns into kinetic energy. Rather more than half passes as heat into his body.

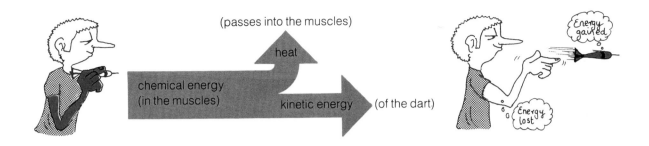

Energy arrows

These fat arrows show what happens to energy when it is used. The arrow heads show how the energy splits into other forms and the thickness of each arrow shows roughly how much energy changes into that form.

2. Catapult

Most of the energy stored in the stretched elastic changes into kinetic energy of the trolley. The wheels warm up a little showing that heat energy has passed into them. Note that this is not the end of the story. The kinetic energy of the trolley will most probably change into other energy forms.

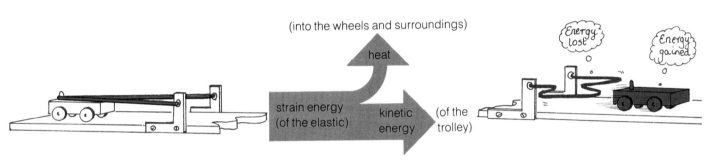

3. High diver

The diver has energy because of her height above the water. Nearly all this energy changes into kinetic energy as she falls. The air is disturbed and warms up a little showing that some of the gravitational energy has changed into heat.

4. Electric motor

Chemical energy in the battery provides electrical energy that makes the motor go round. The weight is lifted off the ground and gains gravitational energy. Chemical energy changes into gravitational energy but passes through electrical energy on the way.

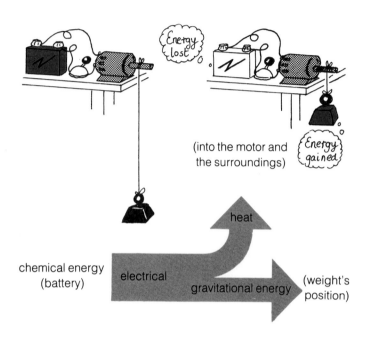

5. Spring-board diver

In this case the beginning and the end of the story are missing. The spring-board diver must have gained his kinetic energy from another form and the energy stored in the bent spring-board will not stay there long before shooting the diver upwards again.

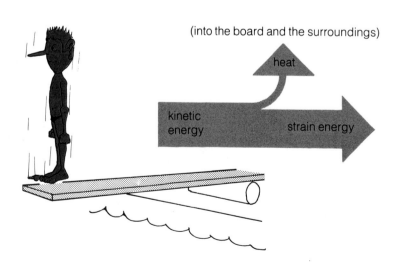
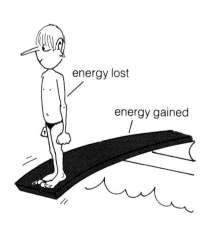

8 At the end of the transfer of energy in the five examples above, objects have lost and gained energy. Copy and complete the table showing the type of energy lost and gained.

Objects	Type of energy	Lost or gained
The muscles of the dart player		
The dart		
The elastic		
The trolley		
The diver on the high-board		
The diver entering the water		
The battery		
The weight		
The spring-board diver		
The spring-board		

9 In all of these examples one other form of energy always appears that is not mentioned in your list above. What form is that?

10 Draw energy diagrams that show the energy changes of the processes that take place in
(a) a radio,
(b) a torch,
(c) a television set (all on),
(d) a fully wound watch.
In each case write a sentence that describes the energy changes that take place.

Where does energy come from?

1. A daily energy supply
Sunlight supplies the Earth with large quantities of energy every day. The energy is free but has to be captured and changed into forms we can use. Here are some of the ways that sunlight is useful:

Photosynthesis
Green plants absorb energy from the sun's light and change it into chemical energy. The plants provide food for humans and animals and wood for cooking and heating.

The sun's day-to-day energy helps provide humans and animals with the food they need

Photocells
These convert the energy of light into electricity. They are used to provide energy for things such as spacecraft, calculators and watches.

Eyes
Light enters the eyes where it is converted into electrical impulses which are then decoded by the brain.

Solar panels providing a satellite with energy

The weather
The Earth's atmosphere absorbs heat from the sun. The heat stirs up the air to make wind and waves. Windmills and wave machines can then be used to change some of the energy of the weather into electricity. The fuel for these machines is free and will not run out.

Evaporation
The sun's heat evaporates water from lakes and seas and lifts it high to form clouds. The water gains gravitational energy from the sun.

After falling as rain its energy can be stored in high lakes held back by dams. The gravitational energy of the water can then be released and converted into electricity by hydroelectric generators.

Meldon Reservoir and Dam, North Devon

Giant windmills being used to drive an electric generator, California

Solar panels
The sun's heat can be trapped by glass. Solar panels and greenhouses use the trapped heat to warm water or rooms.

Direct heating
Everything that lies in the sun is warmed by its rays. Some animals rely on this heating to warm their blood and make them active.

There is enough energy in sunshine for all our needs and we could not live without its daily supply. However it is difficult to convert the energy of the sun into forms that will work machines. Most machines use ancient fuels or electricity made from ancient fuels.

Iguanas basking in the sun, Galapagos Islands

2. Ancient energy supplies

Oil, coal and natural gas supply us with large amounts of stored chemical energy. These fuels were probably formed from trees, plants and small animals that used energy from the sun to grow. They died and were buried and compressed in the ground, eventually forming coal, oil and gas. These ancient 'fossil' fuels took a long time to form and cannot be replaced.

Most machines use either fossil fuel as their source of energy or electricity made from burning fossil fuel. These machines are using up the Earth's limited supply of fossil fuels and alternative sources are needed as reserves begin to run out.

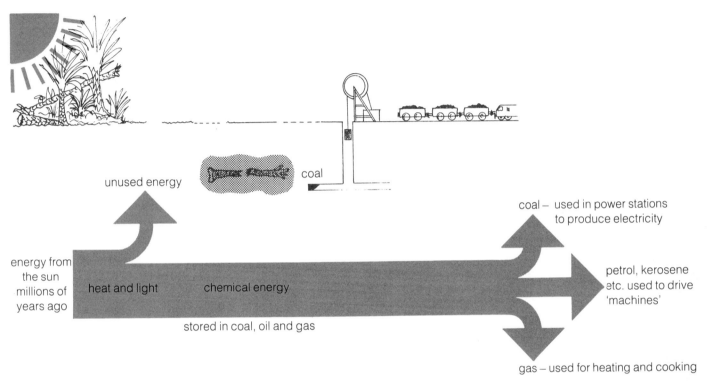

3. Other ancient energy supplies

Radioactive materials in the Earth, mainly uranium, are sources of energy that can be converted into heat and electricity. The supply is limited and there are dangers involved in working with radioactive materials.

The Earth's molten core contains a vast amount of heat energy that sometimes comes to the surface. Water can be used to bring the heat up for our use.

4. Energy from the sun and moon

Tides made by the movement of the sun and moon can be used to turn turbines and make electricity.

Most of the energy we use comes from the sun; from its light and heat. Some was stored as chemical energy long ago and some is supplied daily.

Tides, radioactive materials and the Earth's hot core supply the rest of the energy we use.

Pohute geyser, New Zealand

11 Put this list of the ways we get energy into four columns with these headings:

From sunlight (and heat) today	From sunlight long ago	From the Earth	From movement of the sun and moon

coal, uranium, photosynthesis, solar panels, oil, tides, hydroelectric, photocells, direct heating, gas, geothermal, wind, eyes, gas, waves.

12 Explain the two chief ways in which the sun has provided the Earth with energy.

13 Describe how the sun provides energy to make
   (a) a car move,
   (b) a sailing boat travel along.

14 Write an essay called 'Energy reserves and the future'.

The conservation of energy

Think about a wound-up spring that is accelerating a clockwork toy. Strain energy changes into kinetic energy and heat (the heat energy passes into the toy and makes it slightly warmer). The total amount of kinetic energy and heat energy produced equals the amount of strain energy that disappears.

Energy can be measured, and the unit used to measure all forms of energy is the **joule** (see p. 83). If the spring loses 20 joules of strain energy, then 20 joules of energy must appear in other forms – kinetic and heat in this case. There can be no overall loss of energy. Energy cannot be destroyed.

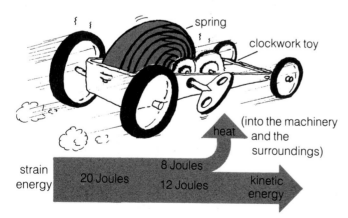

Strain energy changes into kinetic energy and heat. No energy is lost

Energy chains

The beginning and ending of this energy story are missing. The 20 joules of energy stored in the spring came from the person who wound it up. He got it from the food he ate, which got it from the sun. The clockwork toy then dashes across the floor, changing its kinetic energy into heat, warming up itself and the air around. In this way the 20 joules of energy finish up as warmth in the air. Very often, energy chains like this begin with the sun's energy and end up as warmth in the air. But through all the changes no energy is destroyed.

The idea that energy cannot be destroyed and can only change into another form, is called the principle of conservation of energy.

The energy chain for the clockwork is:

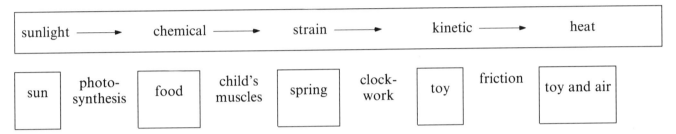

The things in the boxes possess or store the energy as it passes along the chain. The things between the boxes are the converters that change the form of the energy.

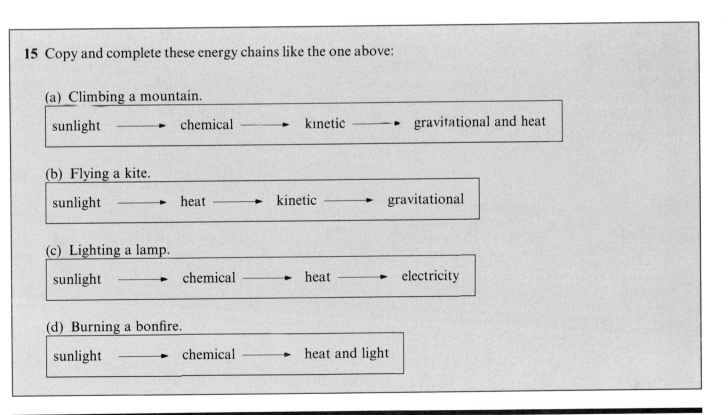

15 Copy and complete these energy chains like the one above:

(a) Climbing a mountain.

sunlight ⟶ chemical ⟶ kinetic ⟶ gravitational and heat

(b) Flying a kite.

sunlight ⟶ heat ⟶ kinetic ⟶ gravitational

(c) Lighting a lamp.

sunlight ⟶ chemical ⟶ heat ⟶ electricity

(d) Burning a bonfire.

sunlight ⟶ chemical ⟶ heat and light

8 Energy transfer

Work

Often when we push or pull an object it moves. When a force moves an object like this we say that work has been done. Energy is transferred whenever work is done.

Examples of doing work
(a) A teacher pulling a boy along on a rope.
(b) Cutting a piece of cheese.
(c) Striking a match.
(d) Going up in a lift.

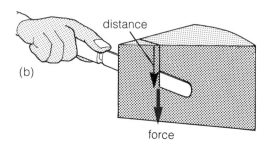

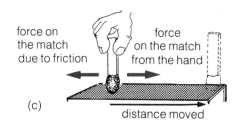

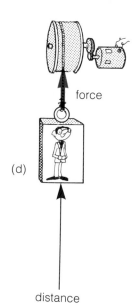

In these examples a force is exerted by an 'engine' that has a supply of energy (the teacher, a hand and an electric motor). This force is great enough to overcome friction or gravity and objects move. A boy gets pulled through the mud, a match is dragged across sandpaper, people are lifted and a knife moves through cheese. In each case the engine does work and loses energy as a result.

The amount of work done is obtained by multiplying the **force × the distance moved in the direction of the force.**

> 1 Write down the situations from this list where work is done. Name the 'engine' that has the energy to do the work and the energy changes that take place:
> (a) drawing with a pencil;
> (b) a table supporting a book;
> (c) a surfer riding a wave;
> (d) a cyclist free-wheeling down-hill;
> (e) pillars holding up a roof;
> (f) a windsurfer moving across a smooth lake;
> (g) a catherine wheel spinning on a post.

EXPERIMENT. Measuring the work done by a steam engine when it is made to lift a load of 1 N

Fix a thread over a pulley to the load that is to be lifted. Run the engine up to top speed and attach the thread so that the load is lifted. Let the engine lift the load by 1 metre and then cut the thread. For later use find the time that the engine takes to do this work. (If you have not got a steam engine, use an electric motor.)

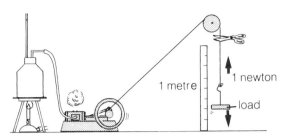

Measuring the work done by an engine

How much work is done

When the engine pulls with a force of 1 newton through a distance of 1 metre, the work it has done is

1 newton × 1 metre = 1 newton metre = 1 joule (joule is the same as 'newton metre').

Work is measured in **joules**.

2 Calculate the useful work done in the examples on p. 82. The forces and distances are given in the table below.

Example	Force (N)	Distance moved	Work done
(a)	500	2 m	
(b)	25	5 cm	
(c)	1	1 cm	
(d)	1000	3 m	

Energy and work

The steam engine in the last experiment changes chemical energy from its fuel into waste heat and gravitational energy of the load. The useful work done by the engine (force exerted × distance lifted) is 1 joule. This work tells us that 1 joule of chemical energy has changed into gravitational energy. The amount of chemical energy supplied by the fuel is much greater than this – it could be 100 J or more – because the engine changes a lot of chemical energy into 'waste' heat.

Note that when an engine works only against friction or wind resistance, all of the chemical energy used finishes up as waste heat. A car travelling along a level road at a steady speed is an example of this.

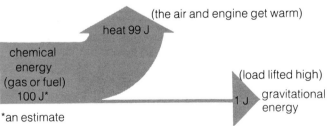

The energy changes that take place as a steam engine lifts a load

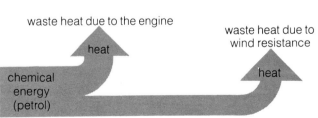

The energy changes that take place when a car travels at a steady speed on a level road

EXPERIMENT. Measuring the work done by an arm

Do 'push-ups' with a 50 N weight. Measure the distance that the weight moves, and for later use, the time it takes to do 20 push-ups. Calculate the work done for the 20 push-ups.

> 3 How much work is done for 1 push-up when a 50 N weight is pushed up 0.8 m? How much work is done for 20 push-ups?

Measuring the work done by an arm

EXPERIMENT. Measuring the work done by a body climbing stairs

When you run up stairs the weight you are lifting is your body weight. So weigh yourself on scales – ones that give a reading in newtons. Then measure the distance from the bottom of the stairs to the top. Run up the stairs as fast as you can. (Note the time it takes to do this for later use.) Calculate the work you have done . . . body weight × distance from bottom to top.

> 4 How much work is done when a boy who weighs 500 N runs up 100 stairs, if each step is 0.5 m in height?

Measuring the work done by a body

Working against gravity

Work has to be done to lift any body above ground level. A force must be used to overcome the pull of gravity and move the body upwards.

The gymnast of mass 50 kg is attracted to the Earth by a force of 500 N ($50 \times g$). To reach the top of the rope she has to pull with a force of 500 N for 6 m. So she must do 3000 J of work to reach the top and her efforts will increase the gravitational potential energy of her body by 3000 J. This is stored energy that can be recovered in other forms. If she lets go and falls, her body will speed up and gain very nearly 3000 J of kinetic energy before it hits the ground. She then has to find a way to stop safely and change 3000 J of energy into heat.

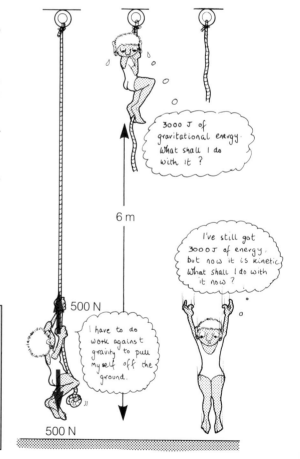

> 4 Describe how the energy changes during the exercise. The forms of energy are: gravitational, kinetic, chemical, heat into the body and rope, kinetic energy of the air, heat into joints and muscles.
>
> 5 The girl loses her nerve and slides down the rope. Describe what happens to the energy in this case.

Two energy equations

Going up

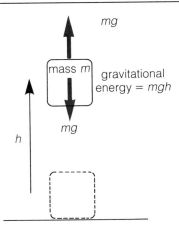

The pull of the Earth on the mass m is mg.
To raise it from the ground requires a force of mg.
The work done to lift it by a height h is mgh.
So the gravitational energy gained $= mgh$.

$$\text{gravitational energy} = mgh$$

This energy depends only on mass and height above the ground and not on how the mass was moved to that height.

Going down

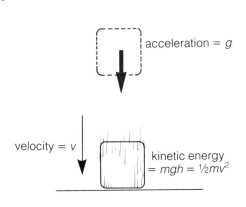

If the mass is released let it take t seconds to reach the ground and reach a velocity of v.
The kinetic energy it gains $= mgh$
but $h = \frac{1}{2}gt^2$ (from $s = \frac{1}{2}at^2$)
and kinetic energy $= mg\frac{1}{2}gt^2 = \frac{1}{2}m(gt)^2$.
Also $v = at$, so $v = gt$
and kinetic energy $= \frac{1}{2}mv^2$.

$$\text{Kinetic energy} = \frac{1}{2}mv^2$$

This equation gives the kinetic energy of a mass m travelling at velocity v, however it acquired that velocity.

---

6 Calculate the gravitational energy of these structures built by a child from six wooden blocks. Each block is a cube with a side of 4 cm and mass of 200 g. ($g = 10$ m/s^2)

7 How much work must a 5 kg monkey do against gravity to reach coconuts that are growing 18 m above the ground? If he grabs a nut and falls what will be his kinetic energy and speed just before hitting the ground? ($g = 10$ m/s^2)

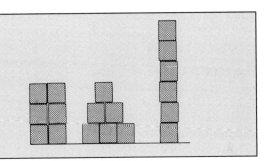

Energy changes

The boy on p. 84 does 25 000 J (25 kJ) of useful work running up stairs. This shows that 25 kJ of chemical energy have changed into gravitational energy. The amount of chemical energy drawn from his muscles is greater than this since some chemical energy is changed into 'waste' body heat.

*an estimate

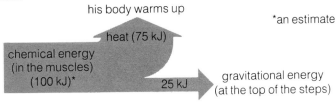

The energy changes that take place as a boy climbs stairs

9 Power

In the last two experiments, your friends may have done the same work as you in less time. If so, they did work faster than you and are more **powerful**.

A powerful engine or person is able to do work quickly. Power is defined as the work done per second. Power can be calculated from this equation:

$$\text{power} = \frac{\text{work done}}{\text{time taken}}$$

Example

If 800 joules of work are done in 10 seconds by a person doing push-ups with a weight, the power of that person's arm is $\frac{800}{10} = 80$ joules/second = 80 watts (watts is short for joules/second). Go back over the last three experiments and calculate the power of the steam engine, your arm and your body. Divide the work done in those experiments by the time it took to do that work.

1 Copy this table and calculate the work done and power of the 'engines' from the figures given.

Engines	Force used (N)	Distance moved (m)	Work done (J)	Time taken (s)	Power in watts (W)
Steam engine	2	1		4	
Arm	40	15		10	
Body	600	10		6	

Power in action

EXPERIMENT. Finding the power needed (in watts) to keep a bicycle moving along a level road

Use a rope to pull a bicycle and rider at 'cycling pace' and a spring balance to measure the force needed to keep them moving. Find the time it takes to travel 10 m from a flying start. The work done is the pulling force (in N) × the distance (10 m). The power is this work divided by the time. Calculate the power that has to be put into the bicycle to keep it going at 'cycling pace'.

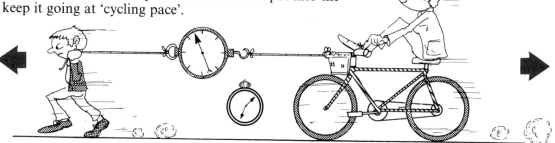

Measuring the power used in cycling

2 In an experiment like this a force of 40 N was used and it took 2 s to cover 10 m. Calculate the power needed to keep the bicycle moving.

EXPERIMENT. Measuring the power of a bunsen burner

This experiment gives a rough value of the (heating) power of a bunsen burner. How much do you think it will be? Use water to collect and measure the energy from the bunsen. Measure the temperature of 1 kg of cold water in a large tin. Put a roaring bunsen under it for 3 minutes. Remove the bunsen, stir and measure the temperature of the water. Work out how much the temperature has risen. It is found that it takes 4200 J of heat energy to warm 1 kg of water by 1°C. Use this figure to calculate the heat supplied by the bunsen. (This energy has been changed into heat by the bunsen and is the 'work done'.) Calculate the heat supplied by the bunsen in 1 second – its power.

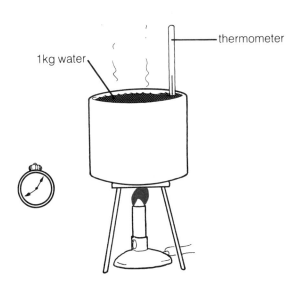

Measuring the power of a bunsen burner

Example

In an experiment like this the temperature of the water rose by 30°C. So the heat supplied by the bunsen was

$$4200 \times 30 = 126\,000 \text{ J in 3 minutes (180 s)}$$

and the power of the bunsen $= \dfrac{126\,000}{180} = 700$ watts.

Note that no force acts in this example so the work done cannot be calculated. In cases like this, power is calculated from energy changed/time taken (p. 277).

EXPERIMENT. The power of an electric motor

Make an electric motor do work by lifting newton weights vertically upwards. The work done on the weights is the number of newtons × the distance lifted (p. 82).

Measure this distance and the average time taken for the lift. Calculate the power of the motor by dividing the work done by the time taken.

Ideas for further investigations:

(a) You could measure the power of the motor for different loads, to see if it varies with motor speed.

(b) You could put an ammeter and voltmeter into the motor circuit to measure the power that the motor takes from the voltage supply (power = voltage × current, see p. 279). This can then be compared with the motor's lifting power.

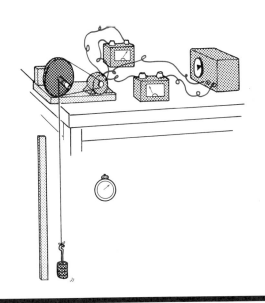

Power problems

In these problems 'engines' do work and convert energy at various rates. Cover the solutions and use the equations

$$\text{power} = \frac{\text{work done}}{\text{time taken}} = \frac{\text{energy changed}}{\text{time taken}}$$

to find the answers if you can. Assume $g = 10$ m/s^2 in all questions.

3 The builder's petrol engine can lift 750 kg of bricks through 7.5 m in 15 s. Calculate
   (a) the gravitational energy gained by the bricks;
   (b) the work done on the bricks;
   (c) the power of the engine in watts.

(a) Gravitational energy = mgh
    = 750 × 10 × 7.5
    = 56 250 J.
(b) Work done on bricks also
    = 56 250 J.
(c) Power of engine
    = $\frac{56\,250}{15}$ = 3750 W.

4 A cyclist, travelling at 5 m/s on a level road, has to overcome a constant force of 50 N due to friction and wind resistance.
   (a) How much work does he do against this force in one second?
   (b) What is his working power at this speed?
   (c) How much energy will he use on a level journey of 18 km.

(a) In one second he will travel 5 m. Work done in one second
    = force × distance
    = 50 × 5 = 250 J.
(b) Working power = 250 W.
(c) Journey time = distance/speed
    = $\frac{18\,000}{5}$ = 3600 s.
    Energy used = power × time
    = 250 × 3600 = 900 kJ.

5 Hadija Muyombo comes from a flat and dry part of Tanzania. She has to walk 3 km to a well for water twice a day. The journey takes half an hour each way. On the way there she works at an average rate of 150 W but on the way back, with full water pots, she has to work at an average rate of 300 W. Her average daily intake of energy from food is 8100 kJ. What proportion of this is used in fetching water?

Energy used on the way = power × time
= 150 × 30 × 60 = 270 000 J.
Energy used on the way back =
twice this. So total energy
used per journey = 540 + 270 = 810 kJ.
For two journeys energy = 1620 kJ.
Fraction of daily intake
= $\frac{1620}{8100}$ = 0.2 = 20%.

6 A bullock has to exert a force of 600 N to drag a plough through the soil. If the animal can cut a furrow 100 m long in 2 minutes, what is its average output power?

Work done in 2 minutes
= force × distance
= 600 × 100 = 60 000 J.
Power = work done/time taken
= $\frac{60\,000}{120}$ = 500 W.

7 This machine was invented by a student to measure the power of human legs. It has a wheel of circumference 2 m that is turned by the legs against a friction belt. There is a small tensioning weight to hold the belt on the wheel and a spring balance to measure the force of friction between the belt and the wheel. Strong legs were able to make the wheel go round 72 times in 24 s with an average frictional force of 60 N and a tensioning force of 4 N. Calculate the power of those legs.

Distance moved against the force of friction in 24 s
= 72 × 2 = 144 m.
Net force of friction
= 60 − 4 = 56 N.
Work done in 24 s = force × distance
= 56 × 144.
Power = work done/time taken
= $\frac{144 \times 56}{24}$ = 336 W.

8 (a) When emptying a bath the water falls 4 m and flows into the drain at 0.8 kg/s. What is the power of the water as it enters the drain? Could it provide a useful source of energy?

(b) In a 'pumped-storage system', water falls 1.8 km from a high lake and enters turbines at a rate of 2 tonnes/second. What power has the 'gravity engine' given to the water?

(a) In each second 0.8 kg of water falls 4 m. So the kinetic energy gained by the water each second = mgh
= 0.8 × 10 × 4 = 32 J,
and the power of this water = 32 W.

(b) In this case 2000 kg of water falls 1800 m each second. So the energy gained by the water from the pull of gravity = mgh
= 2000 × 10 × 1800 = 36 MJ/s
(1 MJ = 1 megajoule = 1 000 000 J)
and the power = 36 MW.

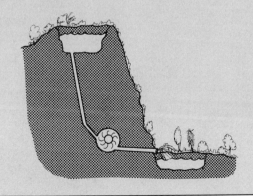

89

10 Machinery

We often use machinery to increase the force we can exert with our muscles. Here are some examples.

Levers

The man is trying to lift a load using a lever. He has arranged the bar and the brick to make the force he has to use (his effort) as small as possible. The lever is a simple machine.

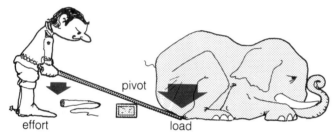

Using a lever to lift a load

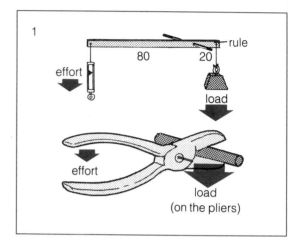

Examples of levers

1. A lever can be used with the pivot between the load and the effort. If the pivot is near to the load, a small effort can lift a big load. Pliers and crow-bars are examples of this type of lever.

2. Some levers have the effort and the load on the same side of the pivot. If the pivot is near to the load, a small effort can lift a larger load.

A wheel-barrow is an example of this type of lever. The pivot is the axle of the wheel; the load is placed as close as possible to the axle, and the lifting effort is put on the ends of the handles.

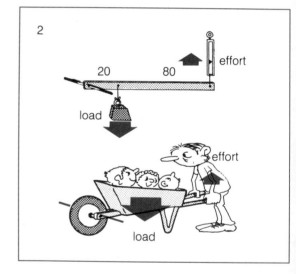

3. Our elbow and fore-arm are an example of another lever arrangement. The load and effort are on the same side of the pivot again but this time the effort is closer to the pivot than the load. It takes more effort to lift a load with this machine than without it.

The advantage with our fore-arm is that a small contraction of the biceps muscle can move the hand through a large distance.

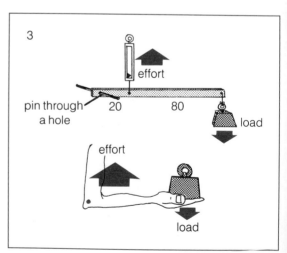

1 Think about the positions of the load, pivot and effort of these levers and say whether they are of type 1, 2 or 3.

Wheel and axle machines

With this type of machine the effort makes the machine go round. The load that is being lifted or turned is attached to a part of the machine that has a small radius. In this way a small effort can lift a larger load. However you will notice that the effort moves through a greater distance than the load.

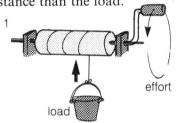

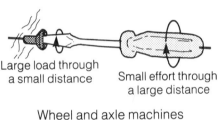

Wheel and axle machines

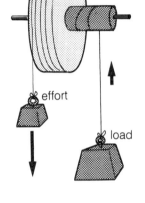

The efficiency of machines

Think about the energy changes that take place when the wheel and axle machine lifts the load shown. The effort loses 100 J of gravitational energy as it falls. We know this because it does 100 J of work (work = force × distance; $50 \times 2 = 100$ J). The load gains 90 J of gravitational energy as it rises. We know this because 90 J of work are done on it ($90 \times 1 = 90$ J). 10 J of energy are wasted and are used to turn the machine against friction.

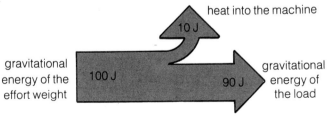

The ratio of energy gained by the load/energy lost by the effort is called the **efficiency** of the machine. Also

$$\text{efficiency} = \frac{\text{work done on the load}}{\text{work done by the effort}}$$

In this example the efficiency $= 90/100 = 0.9 = 90\%$.

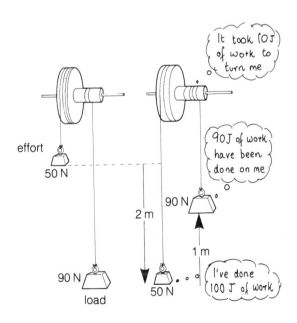

Work being done with a wheel and axle machine

2 This question refers to the wheel and axle machines discussed above. Copy the table and complete.

	Machine 1	Machine 2	Screwdriver
Effort	10 N	4 N	5 N
Distance moved by the effort	100 cm	30 cm	10 cm
Work done by the effort			
Load	45 N	10 N	90 N
Distance moved by the load	20 cm	10 cm	0.5 cm
Work done on the load			
Efficiency of the machine			

◐ EXPERIMENT

Make a wheel and axle machine that can lift a large weight with a smaller one. Measure the work done by the effort and the work done on the load. Then calculate the efficiency of your machine.

3 These diagrams show different wheel and axle machines connected to the same load. Which of the machines will lift the load as the effort moves down? Which machine uses the smallest effort to lift the load.

4 Which of these machines are wheel and axle machines?

Gears and pulleys

Gears are used to pass turning forces from one part of a machine to another. The force and speed that is passed on depends on the sizes of the gears. A large gear always moves more slowly than a smaller gear that is driving it, but can exert more force. If a gear has 10 teeth and drives a gear with 20 teeth, the large gear will go round half as fast as the smaller gear but will be able to exert almost twice the force.

Pulleys connected by belts act in the same way.

5 In each case say whether the pulley that is being driven (white) goes round faster or slower than the driving pulley (black). Also say whether the white pulley goes clockwise or anticlockwise.

6 This diagram shows a 'two-speed' gear box. In which case will the shaft X move slowest, (a) or (b)? This is bottom gear.

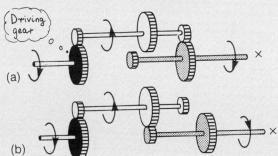

7 In each case say whether gear X moves clockwise or anticlockwise.

8 (a) Does the wheel of a normal bicycle make more turns per second than the pedals?
  (b) If the bike has two chain wheels of different sizes and five gears:
    (i) how many gear ratios can be chosen,
    (ii) which chain wheel and gear would give the lowest gear (for climbing hills),
    (iii) which would give the highest gear (for speed on the flat)?

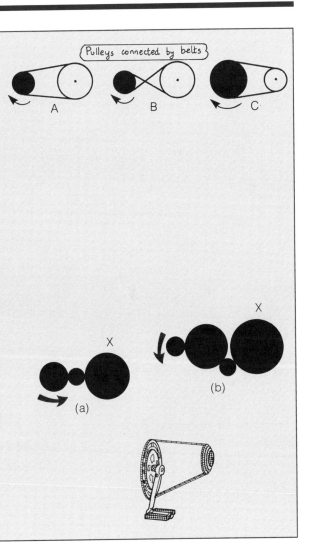

Pulleys for lifting

A pulley is useful for lifting things. It is easier to pull down than to pull up and you can use your weight to help you. Less effort is needed to lift a load if you use two pulleys like this. As rope is pulled from the top pulley wheel, the load and the bottom pulley wheel are lifted. If two metres of rope are pulled out the load will only go up one metre, but it will only take about half the effort. (There are two ropes holding the bottom wheel and they both have to shorten.)

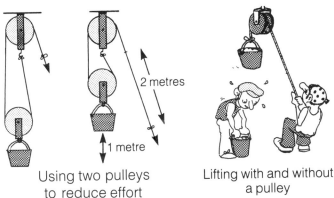

Using two pulleys to reduce effort

Lifting with and without a pulley

9 Copy and complete.

	Pulley (a)	Pulley (b)	Pulley (c)
Load (N)	24	24	24
Distance that the load is lifted (m)	1	1	1
Effort (N)	16	12	8
Distance that the effort has to move			
Efficiency			

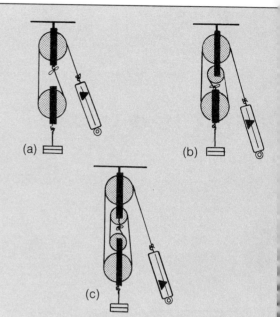

Two, three and four wheel pulley systems

The efficiency of engines

All of the machines studied so far are worked by human energy. They are ways of using human energy to do useful work.

More complicated machines that use other fuels also do useful work for us. These engines use the energy of the fuel to do the work. Not all of the energy is usefully used because some is always 'wasted' as heat. The work done by these engines tells us how much of the energy has been used usefully.

$$\text{efficiency of an engine} = \frac{\text{useful energy got out}}{\text{total energy put in}}$$

$$\text{and also} = \frac{\text{useful work done}}{\text{total energy put in}}$$

10 The four 'heat engines' in this table use energy from a fuel to do useful work. They turn wheels against forces of friction. Calculate the efficiency of each engine and name the fuel that it uses.

	Fuel used	Energy put in per second	Work done per second	Efficiency
Car engine		140 kJ	42 kJ	
Steam turbine		120 kJ	30 kJ	
Steam train		70 MJ	3.5 MJ	
Push bike		600 J	480 J	

11 Wave energy

What is a wave?

EXPERIMENT. Travelling vibrations
Fix one end of a rope and move the free end up and down. A hump is formed that travels along the rope, carrying energy from the hand with it. It is clear that the rope does not move along, but moves up and down as the hump passes. This moving disturbance of the rope is called a wave. When the wave reaches the wall it is reflected but some of its energy is taken by the wall.

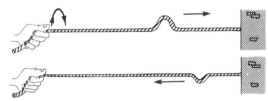

A moving disturbance (wave) on a rope

1 Give one reason why the hump is smaller after reflection.

2 Write down three changes that happen to the hump when it is reflected.

3 Which of the following will make the hump travel faster along the rope?
 (a) moving the rope up and down faster;
 (b) pulling the rope and making it tighter;
 (c) making the rope slacker.

Waves carry energy

Drop a stick into still water and a circular ripple will spread out from the stick. A duck on the water will move up and down as the ripple passes. Continue to move the stick and ripple will follow ripple making a series of humps and dips in the water surface, that travel across the water. These are travelling water waves. The distance between humps (or crests) is called the **wavelength** of these waves.

Travelling waves carry energy

4 What proof is there that water waves carry energy?

5 Why does the duck get so little of the energy you gave to the stick?

6 From where do ocean waves get their energy?

7 Give an example of what the energy of ocean waves can do.

Waves carry vibrations

To make wave follow wave, the bar in the diagram has to be vibrated up and down. The waves carry these vibrations from place to place.

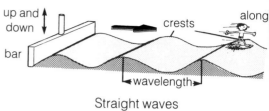

Straight waves

Drawing pictures of waves

A wave is the movement of a disturbance through a material and so changes its appearance as time goes by. This makes it difficult to draw on paper. The pictures show the positions of a wave as it moves along a rope. Each picture shows the wave frozen in time, like a photograph.

The amount the rope has moved up or down from its central position is called its **displacement** and so these pictures can be called 'displacement/distance-along-the-rope' graphs.

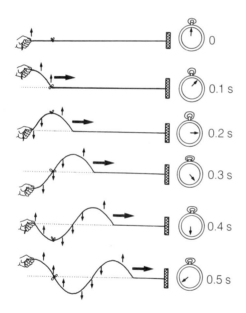

> 8 (a) When time = 0.5 s how many places along the rope have zero displacement?
> (b) When time = 0.5 s how many wavelengths are there on the rope?
> (c) When time = 0.4 s how many places on the rope have maximum displacement?
> (d) When time = 0.3 s is the straight part of the rope being pulled down or up by the approaching wave?

9 The following numbers give the displacement of a rope at different distances along the rope at a particular moment in time.

Displacement (cm)	y-axis	0	0	0	−1.4	−1.3	0.3	2.0	3.0	2.0	0	0
Distance (cm)	x-axis	0	1	2	3	4	5	6	7	8	9	10

(a) Plot a displacement/distance graph of the rope.
(b) How far along the rope does the greatest displacement occur?
(c) If the wave travels along the rope at 10 cm/s show the position of the wave 0.1 s later. Plot the wave's new position on the graph paper using a different colour.

The wave equation

This useful equation can be used to calculate the speed of a wave from its wavelength and the number of waves that pass per second (the frequency). The wave equation is:

$$\text{speed} = \text{frequency} \times \text{wavelength}.$$

These quantities are defined as follows:

The **speed** of a wave is the distance travelled by a crest per second. (v)

The **frequency** of a wave is the number of wave crests that pass an observer per second. The unit is the hertz or Hz. (f)

The **wavelength** of a wave is the distance from one crest to the next. (λ)

Using the symbols, the equation is

v	$=$	f	$\times$	λ
metres/second		hertz		metres

The wave equation can be used for any type of wave.

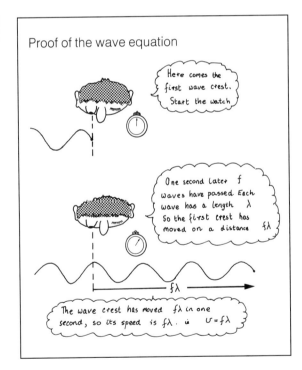

Proof of the wave equation

10 Use the wave equation to calculate the blanks in this table.

	Speed, v (m/s)	Frequency, f (Hz)	Wavelength, λ
Deep ocean wave		0.2	5 m
Wave on heavy rope		1	3 m
Shallow water wave		2	5 cm
Sound wave	330	30	
Light wave	3×10^8		6×10^{-7} m
Microwave		10^{10}	3 cm
Radio wave	3×10^8	10^6	

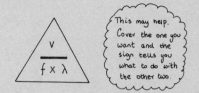

This may help. Cover the one you want and the sign tells you what to do with the other two.

What else can waves do?

You can see how waves behave in your bath but a simple wave-making machine is better for detailed study.

The ripple tank

A ripple tank is used to observe the behaviour of water waves. It has a shallow tray with a glass bottom and shelving sides. (These reduce reflections from the edges.) Water is placed in the tray and lit from above with a small lamp. Circular waves can be made by touching the water with the point of a pencil or by a ball on a vibrating bar; straight waves can be made by dipping the bar in and out of the water. The waves form rather blurred images on a white screen placed underneath the tray and are best seen by looking directly at this screen. A hand stroboscope can be used to 'freeze' the motion of the waves so that it is easier to see what happens when they hit barriers or pass through gaps.

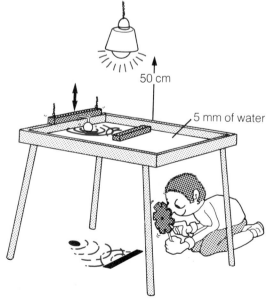

EXPERIMENT. Watching waves

The table describes some important wave properties. See if you can use a ripple tank to make waves do what the diagrams show. In each case check for yourself the observations that are in bold type.

Description	Diagram	Observations
The reflection of straight waves by a barrier.		**The angle of reflection = the angle of incidence.**
The reflection of circular waves by a straight barrier.		The reflected waves are **circular** and have a **centre that is an equal distance behind the barrier.** This is a virtual image of the source (p. 121).
The refraction of straight waves.		The waves **change direction, bending away from the normal,** when they pass from shallow to deeper water. This change in direction occurs because the waves speed up when they pass into the deeper water.
The diffraction of straight waves by a small gap.		The straight waves **become curved** and **spread beyond** the edges of the gap. If the gap is made smaller the waves **curve more, spread more** and **are less high.**

EXPERIMENT. Overlapping water waves

Use a double dipper to tap out two circular water waves in a ripple tank. Make sure that the two waves overlap as they spread out and look at the water that is disturbed by both waves (see photograph).

The overlapping waves can be seen to have two effects on the water. In some areas the waves disappear leaving the water calm. These regions spread like fingers from the two dippers and are places where the effect of one wave has been cancelled by the other. In between these calm fingers the waves have not cancelled but combined to produce a strong disturbance of the water. The pattern is made by the **interference** of water waves.

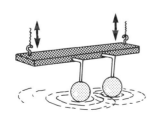

Two circular waves overlapping in a ripple tank

Electromagnetic waves

Some waves are carried by material things such as rope, water and air. There is another sort of wave that can cross space where there is no material. It is called an **electromagnetic** wave. The space around us is full of electromagnetic waves of different wavelengths, coming at us from all directions. These waves carry energy that they deliver to suitable receivers. Waves of certain wavelengths enable us to see. Electromagnetic waves of these wavelengths are called **light.** Waves with other lengths are especially good at warming the skin (called **infrared**) or turning the skin brown (called **ultraviolet**). Some wavelengths can pass right through the body (**X-rays** and **gamma rays**) and others need a radio or television set to receive them (**radio waves**). All this radiation travels at the same speed – the speed of light. It is the length of the wave that determines the effect it has on materials in its path.

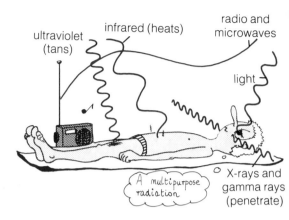

11 Use the sketch above to list the named radiations in order of wavelength.

Microwaves

Microwaves are electromagnetic waves with a wavelength of a few centimetres. They can be made in the laboratory by a high-frequency transmitter and formed into a beam by a specially shaped horn. Microwaves are silent and so a detector with a similar horn is needed to receive them. The detector usually has a meter that gives a measure of the amount of energy carried by the microwave beam.

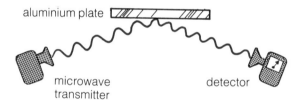

EXPERIMENT. The properties of microwaves

Use a transmitter and detector to check some of the properties of microwaves. If possible use the meter on the detector to estimate the amount of energy picked up in each case. Copy the drawings and write your energy estimates on them.

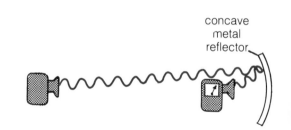

Reflection

Beam microwaves onto a flat metal sheet and search for the reflected beam with the detector. Use a large curved metal dish to collect microwaves and reflect them to a focus. Find this focus with the detector and measure the energy collected by the dish.

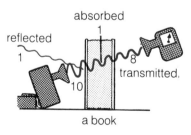

Reflection, absorption and transmission of microwaves (the numbers give an idea of what happens to the wave energy)

Transmission and absorption

Pass microwaves into a book. Use the detector to find out how much energy gets through the book. Also try to measure how much energy is reflected and how much is absorbed by the book.

Pass microwaves through a fine spray of water. Find out if microwaves are absorbed by water drops (and rain).

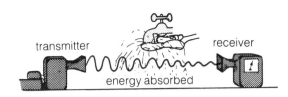

Pass microwaves through a gap between two aluminium plates. Move the detector on the other side of the plates and find how far the beam has spread. The detector should receive microwave energy far beyond the edges of the gap. This spreading of wave energy is called **diffraction**.

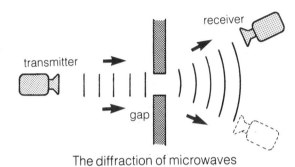

The diffraction of microwaves

12 This diagram shows what can happen to microwaves (and other electromagnetic waves) when they strike matter. They can pass through, be reflected, or absorbed. Copy the diagram and put these phrases into the correct boxes:
(a) 'pass through', (b) 'reflected', (c) 'absorbed'.

13 What happens to the microwave energy that is absorbed?

14 Describe how to show that microwaves can be (a) reflected, (b) diffracted.

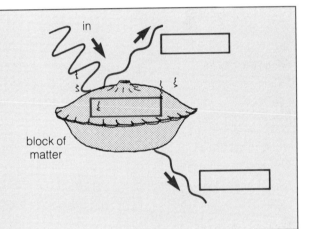

Uses of microwaves

1. Intercity telephone links.
A microwave beam is used to carry telephone conversations between cities. The beam can carry many more conversations than a wire.

2. Satellite links.
A microwave beam, carrying television pictures, can be aimed at a satellite positioned between two continents. The satellite receives and amplifies the beam before sending it back to receiving aerials on the ground. 'Live', world-wide TV coverage is achieved in this way.

3. Cooking.
Microwaves are used in special ovens to cook food. They cook food right through very rapidly. They would also cook our brains and internal organs if we got in their way. Microwave energy should be treated with caution.

A microwave station

Radio waves

Electromagnetic waves with wavelengths between a few millimetres and several kilometres can be called radio waves. Radio waves are widely used to carry messages and television pictures around the world. They can also carry speech and instructions to people and equipment in space. They do this silently at the speed of light.

Radio waves are grouped into bands, each band having a special set of uses. Your radio set may be able to 'pick up' from the long, medium and very short wave bands; some receivers can also pick up short wave transmissions from distant countries. Special equipment is needed to transmit and receive radio waves from other bands.

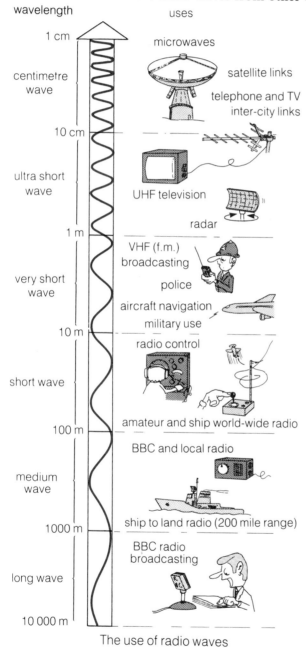

The use of radio waves

A home radio station

15 Write an essay on 'The use of radio waves'.

The electromagnetic spectrum

Radio waves, infrared radiation, light, ultraviolet radiation, X-rays and gamma rays are all wavelike radiations that can pass through the vacuum of space (see also p. 100). They all travel at the same speed in a vacuum – 300 million metres/second – and are electric and magnetic in nature. They form the electromagnetic family of waves. Each type of radiation is produced and detected in its own special way. The wavelengths of these waves vary over a wide range. The diagram shows these radiations spread out according to wavelength into an electromagnetic spectrum.

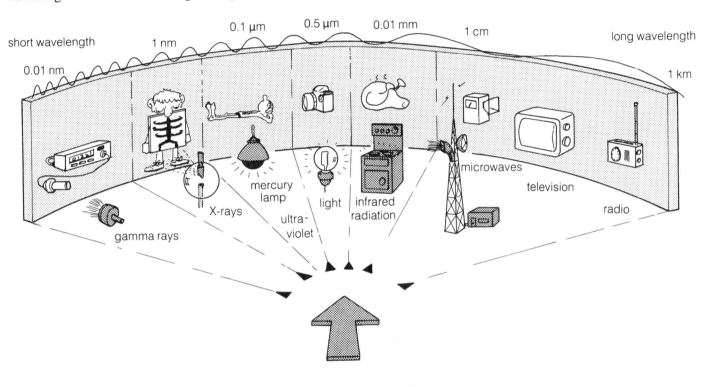

16 Name two other types of wave that cannot pass through a vacuum.

17 Make a list of each type of radiation and what is used to detect it. What does the detector produce when it receives its radiation?

18 Which of these radiations can we detect with our senses?

19 Which type of radiation cannot be switched off?

20 Write down one use that man has found for each type of radiation.

21 Name one type of radiation that
    (a) can pass through a thin sheet of lead,
    (b) causes a suntan,
    (c) can be used to take photographs,
    (d) is used for remote control.

12 Waves we can hear

Vibrations

Backward and forward movements that repeat themselves, can be called **vibrations**. It is difficult to make your hand vibrate more than five times in one second. The prongs of a tuning fork vibrate much more rapidly – so rapidly that it would seem impossible to count them. These vibrations are too fast to see, except as a blur of movement, but they produce sound that we can hear.

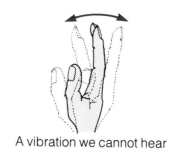

A vibration we cannot hear

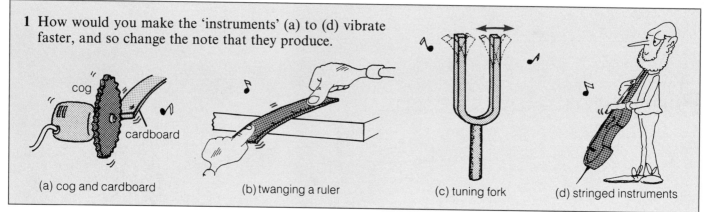

1 How would you make the 'instruments' (a) to (d) vibrate faster, and so change the note that they produce.

(a) cog and cardboard (b) twanging a ruler (c) tuning fork (d) stringed instruments

How many vibrations does a tuning fork make in one second?

EXPERIMENT

Here is a way of showing up the vibrations of a tuning fork so that we can count them. Fix a piece of wire to the prong of the fork and cover a metal disc with an even coat of soot. Put the disc on a record turntable, strike the fork and touch the wire on the revolving disc. As the prong vibrates, the wire will draw a circle of small waves in the soot. Count how many waves there are in a complete circle. Find out how long it takes to draw a circle of waves by finding the time it takes for the disc to go round once. (As this is less than one second, it is best to time 100 revolutions and divide this time by 100.)

Results (get your own if you can).
Time for 100 revolutions = 80 s.
Time for 1 revolution = 80/100 = 8/10 s.
Number of waves in 1 rev. = 200.
Number of waves in one second = $200 \div 8/10 = 250$.

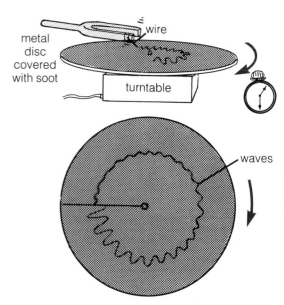

Drawing waves on a sooty disc with a tuning fork

2 Why are the waves at the end of the trace smaller than the ones at the beginning?

3 Explain why the waves are shorter if they are drawn nearer the centre of the disc.

4 The number of waves in a complete circle is the same whether you draw the circle in the centre or near the edge. Explain why this is.

5 Why is the number of vibrations measured in this way likely to be slightly less than the true value? (It is something to do with the wire prong.)

6 A tuning fork scratched a wavy trace in soot on a disc that was rotating on a turntable. The disc took $\frac{1}{2}$ s to go round once and there were 125 waves in one complete circle of waves. What was the frequency of vibration of the tuning fork?

Frequency and hearing

We often need to know how frequent vibrations are. So we call the number of vibrations made in one second the **frequency** of the vibrations. If something makes one complete vibration a second, its frequency is '1 per second' or 1 hertz (1 Hz). For higher frequencies, kilohertz – kHz (or 1000 vibrations per second) – can be used.

A signal generator is an electronic machine that can make a loudspeaker vibrate at different frequencies. The frequency can be read from its control knob.

Using this equipment we find that:
frequencies of less than 20 Hz (20 vibrations per second) are too slow for most human ears to hear as sound;
frequencies of more than 20 kHz (20 000 vibrations per second) are too high for most human ears to hear;
speech is a mixture of vibrations with frequencies up to about 10 kHz. Musical frequencies reach about 16 kHz.

Vibrations with frequencies above 20 kHz produce what is sometimes called ultrasound. This silent vibration has many uses.

7 What is meant by frequency?

8 As the cone of a loudspeaker is made to vibrate at higher and higher frequencies, does the note it gives out get higher or lower?

9 If the cog in question 1 has 25 teeth and the motor revolves eight times a second, what is the frequency of vibration of the cardboard? Would we hear the vibration as sound?

10 Make a rough guess of the following frequencies. You could use the signal generator to match the sounds and get an idea of the frequency that way. Copy and complete the table.

The wings of a pigeon in flight	A fly's wings	The hum of a transformer	The highest note you can make	The lowest note you can make

Sound waves in air

Sound waves carry vibrations from the sound source to our ears and make our ear-drums vibrate.

As the prongs of a tuning fork move out, layers of air are pushed closer together, forming a patch of compressed air. Once formed, this compression moves on through the air until it reaches our ear and pushes the ear-drum inwards. The vibrating prongs send out compression after compression, bombarding the ear-drum and making it vibrate in step with the tuning fork.

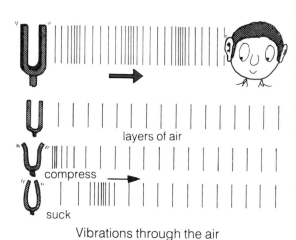

Vibrations through the air

Longitudinal waves

The layers of air behave rather like the links of a spring. If the hand holding the spring is moved to the right, the spring is compressed over a small distance. This 'compression' will then travel along the spring. Further compressions will follow if the hand continues to move backwards and forwards. Such travelling compressions are called a **longitudinal** wave. Sound waves are also longitudinal waves.

11 Which way does one particular coil of the spring move, as the wave passes by?

12 How would you move your hand to make humps travel along the spring? Which way would one particular coil move in this case?

Transverse waves in a spring

The compressions are reflected when they reach the fixed end and travel faster if you stretch the spring

Longitudinal waves in a spring

13 How can you make the wave travel faster along the spring?

14 What is the difference between a longitudinal and transverse wave?

Other ways of making longitudinal waves

Try these experiments:

Push one end of a queue of people. The longitudinal pulse passes along the line to the last person, who wonders what has hit him.

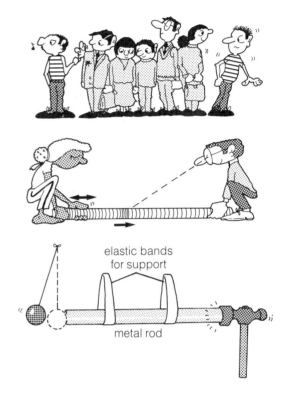

Pluck one end of a long spring. Pull some links towards you and then let them go. A dark region of compressed links passes along the spring and moves the hand of the person at the far end.

Tap a hanging metal rod with a hammer. An unseen longitudinal compression travels along the rod. When the compression reaches the other end, the ball is given a kick that sends it flying. The fast-moving compression then bounces between the ends of the rod and makes it ring.

Note that in all these examples:
(a) The materials 'carry' the waves and do not move far themselves.
(b) The longitudinal wave travels in the same direction as the small vibrations of the materials.
(c) There is no wave hump as with transverse waves.

Sound waves in solids

Try the experiments shown in the pictures.

You will probably find that sound travels well through wood, metal and string. It also travels faster in these materials than in air (3600 m/s for steel).

The vibrations of the noisy object send small longitudinal compressions through the solid. These rapidly reach the far end and transfer the energy of their vibrations to the air or our ears.

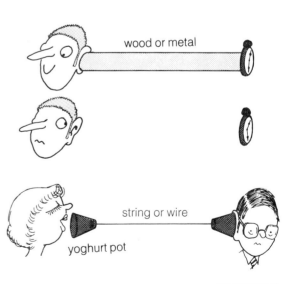

Sound waves in liquids

Next time you go swimming try speaking to a friend under water. It is quite difficult to speak clearly but the sound, once made, travels well through the water. Find out how far apart you can go and still hear each other.

Sound travels well through water and special microphones and loudspeakers have been developed to use underwater sound.

Passive sonar

Underwater sounds made by submarines or even whales can be picked up by underwater microphones (hydrophones). Sound is a powerful way of detecting objects that are hidden from us by water.

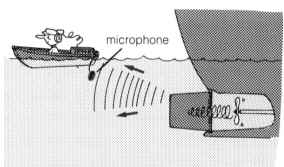

Active sonar

To detect silent objects such as a shoal of fish or a submarine lying on the bottom, sound pulses can be sent out by a ship or buoy. These pulses are reflected by the submarine and picked up by the ship. The echo gives away the position of the submarine.

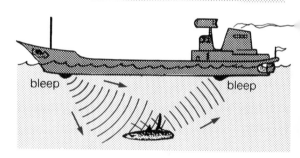

15 What is the difference between active and passive sonar?

16 A ship sends out a pulse of sound underwater and displays both the pulse and its echo from the bottom on an oscilloscope screen. The time between the two is measured as one second. How deep is the water? (Speed of sound in water = 1430 m/s).

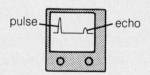

17 List some of the uses of sound under water.

Sound waves in a vacuum

Can sound waves travel through a space when there is nothing there?

EXPERIMENT

Hang a bell (or small battery radio) inside a jar that can be evacuated by a vacuum pump. Switch on the bell and listen as you pump the air out of the jar.

If the sound from the bell dies away, it shows that the sound waves cannot travel through the vacuum you have made. (Some of the vibrations from the bell may reach the outside through its supporting strings.)

18 What could you use to support the bell to reduce the vibrations that get out through the support?

19 How can you tell that the bell is still vibrating in the vacuum?

20 Describe exactly what you hear when air is let back into the jar.

21 Why did American Indians put their ears to the ground when listening for approaching horses?

22 Why are the gurgles and clunks of a central-heating system often heard throughout the house?

23 Give a reason why astronauts use radio waves to talk to each other on the moon.

24 Describe how you would show that sound cannot travel through a vacuum.

25 Give two reasons why you can hear sounds from the **inside** of an engine if you listen to it through a stick.

The speed of sound in air

EXPERIMENT

How fast does sound travel? Here is a way of measuring its speed.

Stand 50 m from a wall that gives a good echo. Make a loud clap and listen for the echo. It takes less than a second for the sound to get back to your ears – too short a time to measure with a stopwatch. So make a second clap the moment the echo is heard. Keep this up, clapping with a rhythm that makes each clap cover the echo of the previous clap. Time 20 of these claps, counting the first clap as '0'.

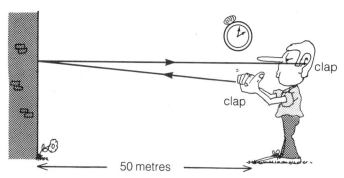

Measuring the speed of sound

Results (get your own if you can).
Time for 20 such claps = 6 s.
Time between clap and echo = 6/20 s.
Distance travelled in this time = 100 m.

Speed of sound = $\dfrac{\text{distance}}{\text{time}}$ = $100 \div \dfrac{6}{20}$ = 333 m/s.

(As the time measurement is only approximate, this result should be given as 330 m/s).

26 Why is it difficult to measure the time between a clap and its echo using a stopwatch?

27 Give two reasons why the echo is quieter than the clap (see p. 95).

28 How far must you stand from a wall if you want the sound of your clap to reach your ears after a journey of 200 m?

29 If the BBC used sound waves instead of radio waves for broadcasting, how long would it take the programme to reach Perth from London (666 km)? (Take the speed of sound as 333 m/s.) What other problems would there be?

The shape of sound waves

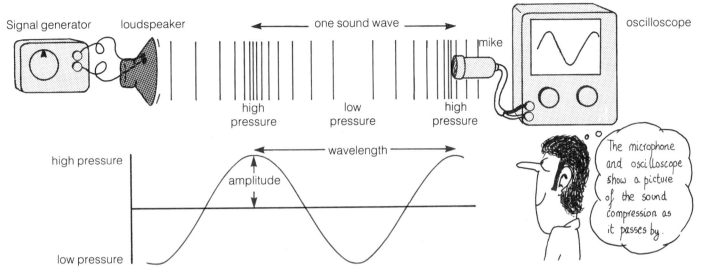

Looking at the shape of sound

The pressure changes through a simple (pure) sound wave look like this. Such a wave has a **wavelength** (distance from crest to crest) and an **amplitude** (height of the wave above the centre line). Using the equipment above we find that:
LOW notes have LONG wavelengths.
HIGH notes have SHORT wavelengths.
LOUD notes have a LARGE amplitude.
SOFT notes have a SMALL amplitude.
Notes from different instruments have different shapes (see diagram). It is the shape of a sound wave that enables us to identify the instrument it came from.

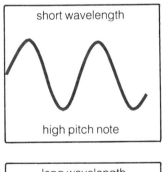
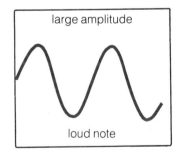
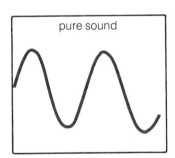
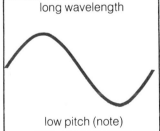
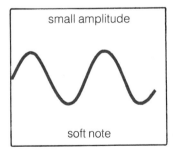
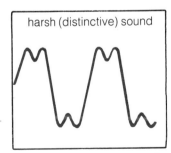

these two notes differ in quality

How changes in a sound alter the shape of its wave

30 Say whether the sound waves shown have the same or different wavelengths and amplitudes. Copy and put same or different.

The two sound waves	Amplitude	Wavelength
(a) (b)		
(c) (d)		

31 Which of the four sounds shown above would be: (a) the softest, (b) the highest note?

32 A note played on a piano sounds different from the same note played on a violin. Explain why this is so.

Energy and Waves check list

After studying the sections on Energy you should:
- know that energy can be stored in different forms
- know that moving bodies have kinetic energy
- know that some forms of energy are only found in transit
- be able to work out real-life examples of how energy changes from one form to another
- know where energy comes from
- be able to trace energy chains for real-life examples
- know the principle of conservation of energy

After studying the section on Energy transfer you should:
- know what is meant by work
- know how to calculate the work done by a force
- know that work is measured in joules
- know how to calculate power and that it is measured in watts
- know how to measure the work done and power of human (and other) machines
- know how to calculate work done by gravity and against gravity
- know how to calculate gravitational potential energy and kinetic energy
- know what is meant by a machine
- know examples of the uses of levers and wheel and axle machines
- know examples of the use of pulleys and gears
- know how to calculate the efficiency of machines

After studying the section on Waves you should:
- be able to describe a transverse wave
- be able to interpret displacement/distance graphs of waves
- know the meaning of amplitude, wavelength and frequency
- know the wave equation and how to use it
- know that waves can be reflected, refracted and diffracted
- be able to name some electromagnetic waves and their effects
- know the audible frequency range
- be able to describe a longitudinal (sound) wave
- know that sound waves can travel in solids, liquids and gases but not in a vacuum
- be able to meausure the speed of sound in air
- know the effect of amplitude, wavelength and frequency on how a wave sounds

Revision quiz

Use these questions to help you to revise the sections on Energy and Waves.

- Name three forms of stored energy.
  ... chemical, gravitational and strain.
- What is kinetic energy?
  ... the energy of a body due to its movement.
- Name three forms of energy only found in transit
  ... light, heat and sound.
- Why is electrical energy so useful?
  ... because it can be changed into so many different forms.
- What form of energy is always produced when energy changes form?
  ... heat energy.
- Name a process that changes sunlight into chemical energy.
  ... photosynthesis.
- Name other ways in which the energy of sunlight can be converted.
  ... by photocells, eyes, weather, evaporation, solar panels, direct heating.
- Name three fossil fuels.
  ... coal, oil and natural gas.
- Why can't fossil fuels be replaced?
  ... because they take millions of years to form.
- Name three non-solar energy sources.
  ... uranium, geothermal and tidal energy.
- What is an energy chain?
  ... it is a chain that follows energy from its original to its final form.
- What is the principle of conservation of energy
  ... this states that energy cannot be created or destroyed.
- When does a force do work (and transfer energy)?
  ... when it moves the object it is acting on.
- How do you calculate the work done by a force?
  ... work done = the force × the distance moved in the direction of the force.
- What is the power of an engine?
  ... it is the work it does per second.
- How is power measured?
  ... power = work done/time taken.
  ... 1 watt = 1 joule/second.
- How do you calculate the work done raising a mass against gravity?
  ... work done = mass × g × height lifted = mgh.
- How do you calculate the gravitational potential energy of a mass?
  ... gravitational potential energy = mgh.
- What is the wavelength of a wave?
  ... the distance from the top of one wave crest to the next.
- What is the frequency of a wave?
  ... the number of waves that pass per second.
- What is the wave equation?
  ... wave speed = frequency × wavelength.
- When is a wave diffracted?
  ... when it passes through a gap a few wavelengths wide.
- What happens to a straight wave when it is diffracted?
  ... it becomes curved and spreads beyond the edges of the gap.
- Name some electromagnetic waves.
  ... gamma rays, X-rays, ultraviolet, visible, infrared, microwaves, radio waves.
- What is the audible frequency range of normal ears?
  ... about 20 Hz to 20 kHz.

- What is a longitudinal sound wave?   ...it is a wave where the displacement is in the direction that the wave travels.
- Can sound travel through space?   ...no, sound must have a solid, liquid or gas to carry its vibrations.
- What determines the loudness of a sound?   ...the amplitude of its wave.
- What determines the pitch of a sound?   ...the frequency of its wave.
- What determines the quality that distinguishes a sound?   ...the shape of its waves.
- How do you calculate the kinetic energy of a body?   ...kinetic energy = $\frac{1}{2} mv^2$.
- What is a machine?   ...it is a device that changes the size or direction of the force applied to a load.
- Give an example of each of the three types of lever.   ...a crow-bar, a wheel barrow, the forearm.
- Give examples of wheel and axle machines.   ...door knob, handle bars, brace and bit, screw driver.
- What is meant by the efficiency of a machine?   ...it is the amount of energy that comes out of it compared with the amount put in.
- Can the efficiency of a machine be more than 100%?   ...no, it is always less than 100%. The useful energy got out is always less than the energy put in.
- How can efficiency be calculated?   ...efficiency = work got out/work put in.
- What is a transverse wave?   ...it is a wave where the displacement is sideways to its direction of movement.
- What is the amplitude of a wave?   ...the distance from the centre to its highest point.

Examination questions

1. Fig. 1 shows the electromagnetic spectrum.
   (a) Copy Fig. 1 and fill in the names of the two missing regions.
   (b) Which region
       (i) has the longest wavelength,
       (ii) has the highest frequency,
       (iii) causes a suntan,
       (iv) is used in burglar alarms?
   (c) Some washing powders contain a chemical which is sensitive to ultraviolet radiation. State and explain what you see when clothes washed in such a powder are put in sunlight.
   (NEA)

Fig. 1

2. Fig. 2 shows a beam of sonar waves sent to a shoal of fish directly underneath a fishing boat.
   (a) The speed of the sonar waves in water is 1400 m/s and the echo returns after 0.1 seconds. Use the equation

   $$\text{distance} = \text{speed} \times \text{time}$$

   to calculate the depth of the shoal of fish (d).
   (b) Explain why the returning pulse lasts for a longer time than the pulse sent out.
   (NEA)

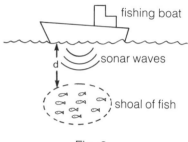

Fig. 2

3. (a) Fig. 3 is an incomplete diagram which shows three successive straight waves A, B, and C on water, as they are being reflected at a straight barrier XY. Wavefront A is just about to be reflected while B and C have already been partly reflected.
   Copy and complete Fig. 3 showing the positions of the reflected parts of the wavefronts B and C.
   (b) Fig. 4 is an incomplete diagram which shows a circular wavefront originating at O just before reflection at a straight barrier AB.
   Copy and complete Fig. 4 by drawing the wavefront as it would be just after reflection is complete. Indicate on the diagram where the reflected wavefront appears to come from.
   (c) If the wavelength of an incident wave is 1.5 cm and the frequency of the source at O is 10 Hz, calculate
       (i) the wavelength of the reflected wave,
       (ii) the speed of the waves over the water.
   (NEA)

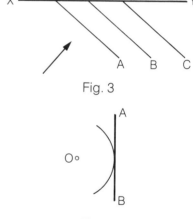

Fig. 3

Fig. 4

4. (a) Waves can be used to carry energy. Some waves are described as being transverse and others are longitudinal. How do transverse waves differ from longitudinal ones?
   (b) Which of the following are transverse waves and which are longitudinal?
   (i) Ripple tank waves on water, (ii) sound waves, (iii) light waves, (iv) radio waves.
   (LEAG)

5 The diagram shows the main regions of the electromagnetic spectrum.
   (a) On a copy of the diagram write in the names of the two blank regions.
   (b) Which radiation shown in the table has the SHORTEST WAVELENGTH?
   (c) Which radiation shown in the table has the HIGHEST FREQUENCY?
   (d) In a vacuum, is the speed of gamma rays greater than, the same as or less than the speed of radio waves?
   (e) State ONE use of gamma rays.
                                                                    (SEG)

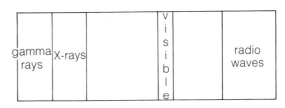

Fig. 5

6 This question is about FORCES, WORK and ENERGY.
   A worker on a building site raises a bucket full of cement at a slow steady speed, using a pulley like that shown in the diagram. The weight of the bucket and cement is 200 newtons. The force F exerted by the worker is 210 newtons.
   (a) Why is F bigger than the weight of the bucket and cement?
   (b) The bucket is raised through a height of 4 metres
       (i) Through what distance does the worker pull the rope?
       (ii) How much work is done on the bucket and cement?
       (iii) What kind of energy is gained by the bucket?
       (iv) How much work is done by the worker?
       (v) Where does the energy used by the worker come from?
                                                                    (SEG)

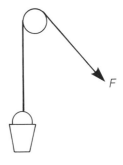

Fig. 6

7 In a hydroelectric generating station water falls through pipes from a high reservoir to a turbine. The turbine drives an alternator.
   (a) Copy the block diagram in Fig. 7 and label it with the three parts of the system. On the diagram show clearly the main energy changes in the system.
   (b) Write down the names of THREE different sources of energy which are used for driving an electrical generating system.
                                                                    (SEG)

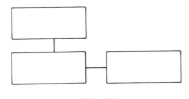

Fig. 7

8 A microphone is connected to a cathode ray oscilloscope. Three sounds are made in turn in front of the microphone. The traces A, B and C produced on the screen are shown in Fig. 8. (The controls of the oscilloscope are not altered during the experiment.)
   (a) Which trace is due to the loudest sound?
       Explain your answer.
   (b) Which trace is due to the sound with the lowest pitch?
       Explain your answer.
                                                                    (MEG)

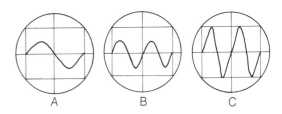

Fig. 8

9 (a) Explain the meaning of the terms work, energy and power, stating the units in which each is measured.
(b) Describe in detail an experiment by which a pupil could measure the average power he or she can develop over a period of a few seconds.
(c) A crane, whose engines can develop a power of 2 kW, is used to raise a 400 kg load from the ground to a platform 12 metres above the ground.
  (i) Determine the work done in raising the load on to the platform.
  (ii) Determine the minimum time for the crane to raise the load on to the platform.
  (iii) Determine the minimum time for the crane to raise a load of 100 kg on to the same platform.
(Take the acceleration of free-fall to be 10 m/s^2.)
(d) State what is meant by the term *conservation of energy*. Discuss how the idea of energy conservation can be applied to the example of the crane in part (c).

(NISEC)

10 (a) A model railway train starts from rest and runs along a straight track about 3 m long. Describe with the aid of a suitable diagram how you would make measurements from which you could work out the speed of the train at different times from the start. Explain carefully how you would calculate the values of the speed.
(b) A motor car of weight 7000 N is travelling along a level road at a constant speed of 20 m/s. When it comes to a hill, which rises vertically 100 m and is 2 km long, the driver increases the power output of his engine to keep the speed constant at 20 m/s.
  (i) How long does it take the car to climb the hill?
  (ii) How much work does the car do against gravity as it climbs the hill?
  (iii) What power is needed to do this work against gravity?
  (iv) On the return journey the car crosses the top of the hill in the opposite direction at 20 m/s and the driver then disconnects the engine by pushing the clutch pedal down. Explain, in terms of energy changes, why the speed of the car increases. Include the effects of air resistance in your explanation.

(MEG)

13 Light energy

Light carries energy

The pictures show things that give out light. They are called luminous bodies. Most things are not luminous and do not give out light. We see them by the light they reflect into our eyes.

A luminous object produces light which travels away from it at a very high speed (300 000 km/s). The light carries energy away from the object as an electromagnetic wave.

Some materials can change the energy of the light into other forms. Photographic film uses light energy to store pictures. Green plants use it to make sugars and starch which store chemical energy (photosynthesis). Eyes use it to form electrical images that are sent to the brain.

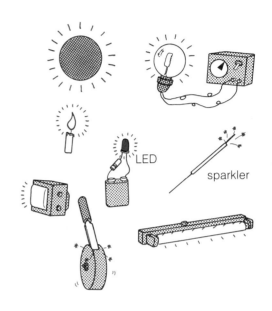

1 Copy this table and say whether the energy used by each luminous body is nuclear, electric, chemical or kinetic. In the second box say whether heat is produced as well as light.

	Sun	Filament lamp	Candle	LED	Sparkler	TV screen	Grind wheel
Energy used							
Lots of heat too?							

2 Make a list of other luminous bodies you know.

Forming light energy into an image

EXPERIMENT. Using a lens to form an image on a screen

Arrange a lamp, lens and screen along a metre rule on the bench, so that a focused image of the lamp is formed on the screen. Move the lens and screen about and make the largest image you can. And then the smallest possible image. Notice that the image is upside down and reversed. The image is a likeness of the lamp, not an exact copy and differs from the lamp in a number of ways.

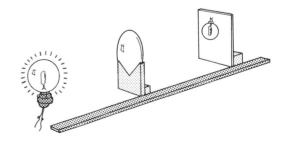

3 (a) Which arrangement will make the largest image?
  (b) What must you do to make the image larger (two things)?

4 Are the following the same or different for the object and its image? Copy and put 'same' or 'different' in the table.

Colour	Brightness	Shape	Way up	Way round

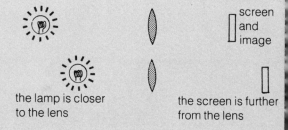

the lamp is closer to the lens

the screen is further from the lens

🌑 EXPERIMENT. A home-made projector

Use a colour slide, lit by a lamp as an object. Place a lens and screen as shown and move them about until you have a large image of the slide on the screen. You will notice that this image is upside down and reversed. Since it is formed on a screen, it is called a real image.

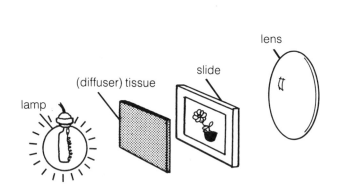

A home-made slide projector

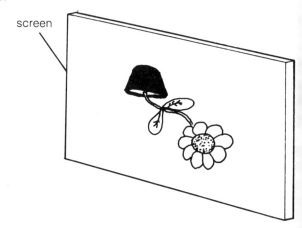

5 Criticize your home-made projector.

Magnifying glass

A lens can form a different sort of image. Hold a lens close to your eye and focus on your thumb. You will see an image that is upright, magnified and the right way round. It is the image we make when we use a magnifying glass to look at small things. This sort of image cannot be shone onto a screen and is called a virtual image.

A lens forming a virtual image

Mirror images

Look into a mirror and wink your left eye. The image winks back with its right eye. The face we see in a mirror is not the face others see when they look at us. The mirror image is reversed, left to right. A photograph, film or television picture shows us as we really are.

6 (a) One of these photographs was printed the wrong way round. How could you make one photograph look like the other? Is the photograph that is printed the wrong way round the same as the mirror image of the person?

(b) How do you think the other two photographs below were made? Are they of the same person?

EXPERIMENT. Finding the mirror image of a candle flame

Place a lighted candle in front of a sheet of glass. Look into the glass and you will see an image of the bright flame. The glass is acting like a mirror. Find the image by moving an unlit candle on the other side of the mirror. Position this candle so that the image of the flame sits on its wick. The unlit candle must now be in the same place as the image. What do you notice about the distances of the two candles from the mirror?

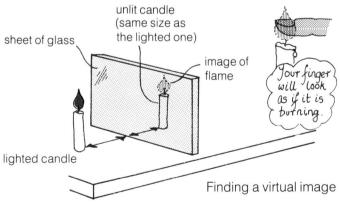

Finding a virtual image

Virtual images

We have seen how a lens forms a real image on a screen. A mirror image can be seen by the eye but does not give a picture on film or on a screen placed where the image is formed. Such an image is called **virtual**.

7 Why is the sign on the front of an ambulance sometimes written as shown below? Would it be sensible to write the sign on the back of the vehicle like this?

ƎƆNAJUBMA

8 How could you arrange a candle, a sheet of glass and a beaker of water to give the illusion of a candle burning in a beaker of water?

9 Explain how this shop window trick works.

10 If you walk towards a mirror at 1 m/s, how fast do you approach your image?

11 A sheet of glass forms two images of a candle flame that are close together. Explain how the glass does this.

12 Which letter in the diagram on the right shows the position of the image of the girl's toes? Explain why she cannot see her hair-do in the mirror.

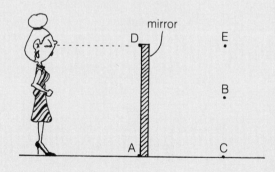

13 This diagram shows the image in a plane mirror of a clock face. What is the correct time: (a) 2.35, (b) 8.35, (c) 9.25, (d) 2.25?

14 Light rays

A narrow beam of light is often called a ray.

Use a ray box or lamp and slit, to produce a long fine ray of light. Experiment with the light by shining it at a mirror, a sheet of glass, a prism, etc. You can assume that rays always travel in a straight line, even over long distances.

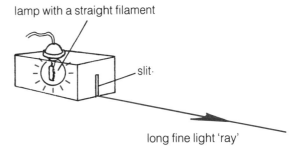

A ray box sending out a light 'ray'

EXPERIMENT. The law of reflection

Look at the way a mirror reflects a ray of light.

Draw two lines at equal angles to a plane (flat) mirror and shine a ray of light along one of the lines. You will see that the mirror reflects the ray along the other line. The ray and its reflection always make the same angle with the mirror. This is the law of reflection.

A line at right angles to the mirror at the point where the ray strikes is called the **normal**. The angles between the rays and the normal are called the angle of incidence and angle of reflection. This law of reflection can be written:

  angle of incidence = angle of reflection

You will also find that the rays and the normal all lie on the same flat surface.

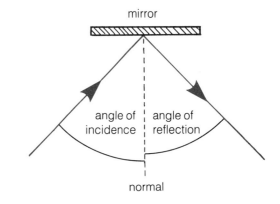

Light levers

If you move the mirror slightly in this experiment, you will see the reflected ray move through a much larger distance. A long light ray reflected from a mirror like this greatly magnifies the movement of the mirror. Such an arrangement is called a light lever.

---

1 Explain how a cup of tea in a sunny room can produce a flickering patch of light on the ceiling.

2 A boy is using a mirror to flash a signal to his friend.
  (a) How does he know when the sun's reflected light rays will be seen by his friend?
  (b) What will his friend then see?
  (c) How can the boy send a series of quick flashes of light to his friend?

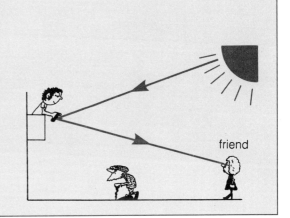

Seeing by scattered light

In question 2 the girl sees the bright image of the sun because she is in the right position to receive the light reflected by the mirror. The old man would not be dazzled by the light. But both people see the boy equally well from different angles. This is because his body scatters light from the sun and sends it out in all directions. The scattered light makes him visible from different directions.

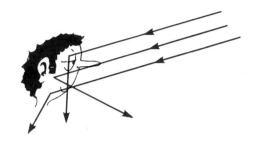

A closer look at mirror images

○ **EXPERIMENT. Using rays to find a mirror's image**

This experiment traces rays of light from a bright object point and looks at the pattern of their reflection.

Draw a line in the middle of a sheet of paper to mark where the mirror will go. Make a dot 'P' on one side of the mirror and draw four lines from the dot to the mirror. Shine a fine strip of light from a ray box along each line and mark the reflected rays so that you can draw them in later.

An image formed by deception

P represents a bright point object, such as the tip of a candle flame, that sends out rays of light towards the mirror. The experiment shows how four such rays spread out after reflection.

Note how the reflected rays appear to have come from a point behind the mirror. You can check this by continuing the reflected rays backwards until they meet at q. (If the rays do not meet at an exact spot, remember that it is impossible to experiment with perfect accuracy.)

The eye – or the lens of a camera – collects the reflected rays and supposes they have come from q. It is not aware that the rays have been folded back and sees an image at q. There is nothing actually at q and so this type of image is called 'virtual'.

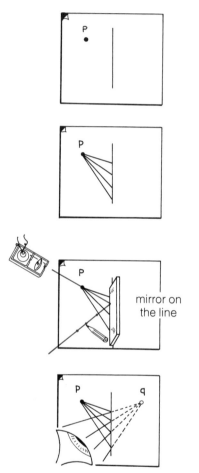

mirror on the line

3 Two rays have been drawn from the FOOT of an arrow that is lying in front of a mirror. They show that an image is formed at TOOᖵ. Draw two similar rays from the head of the arrow to show where the image of the head is formed. Label this image with the word HEAD as it would be seen in the mirror.

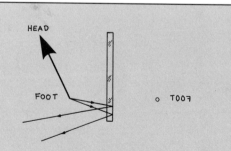

4 Use this diagram of the reflection of a ray from a mirror to prove that if the angle of incidence (i) = the angle of reflection (r), then the distances x and y are equal.

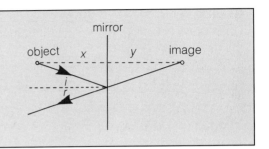

Light rays and concave reflectors

Try these experiments with a concave mirror. See if you can use the mirror to reflect the rays in these ways:

1. Can you reflect several rays to a focus?
Use a ray box with three or more slits to make the rays. Hold the mirror half over the edge of the bench and use a sheet of paper to show up the rays.

2. Can you reflect a broad beam of light (very many light rays!) to a focus? Take away the slits to produce the beam.

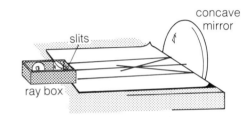

3. Can you reflect the sun's light and heat to a 'hot spot'?

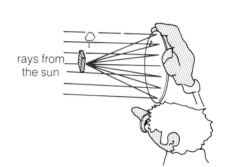

4. Can you reflect light from a small lamp near the mirror into a beam as wide as the mirror?
Move the lamp until the whole mirror is filled with light.

Notice how rays of light can be brought together by reflection in a concave mirror. The energy of the light is collected by the mirror and concentrated into a small region. The light rays are converged by the mirror.

Note also how the same mirror can have the opposite effect. It can reflect the light from a small lamp out into a broad beam.

Other electromagnetic radiations, such as microwaves, infrared and ultraviolet rays, can be reflected in the same way as light. The longer waves (infrared and microwaves) need larger mirrors to collect their energy properly.

5 When a concave mirror is sending out a broad beam of light from a nearby lamp, is it still converging the rays? (Converging means 'making the rays spread less'.)

6 Concave reflectors have many uses. For each of the uses shown decide whether
 (a) the reflector concentrates the energy of a broad beam to a small point, or
 (b) sends out energy from a small source into a broad beam.

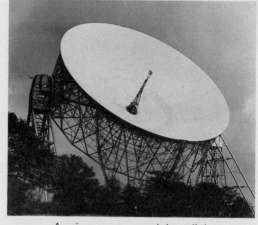
A microwave receiving dish

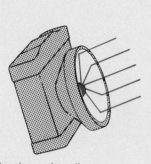
torch, car headlamp, photographic flash gun

reflecting telescope

paraffin heater

solar cooker

The refraction of light

When a ray of light passes from air into a clear material, it may bend sharply. This sudden change in direction is called **refraction**.

Water and other clear liquids refract light. So do all transparent solids such as Perspex and diamond. Even gases refract light but by very small amounts. Refraction occurs because light changes speed when it goes from one material to another. It travels slower in solids than in air and most liquids.

EXPERIMENT. Refraction and reflection of light rays

Pass a ray of light into a block of glass and look for the reflected and refracted rays. Draw round the block and mark the path of all the rays you can see. (The ray inside the glass may not be visible but you can work out where it goes from the rays outside.)

Notice how the ray changes direction when it goes into glass and bends 'towards the normal'. When it leaves the glass it bends the other way and comes out parallel to the way it went in.

Each surface reflects some of the light and refracts the rest. Change the angle of the block and notice how well glass reflects light when the angle is large, and how well it refracts light when the angle is small.

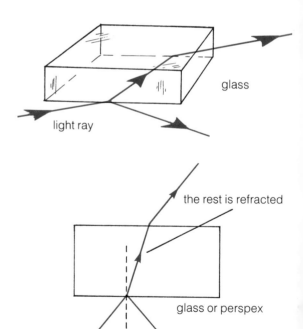

7 Explain the terms reflection and refraction.

8 Copy this picture of light striking a glass of water. Show what happens to the energy of the light by putting the words **transmitted**, **reflected** and **absorbed** into the right boxes.

9 Copy the picture below and continue the light ray into the air and the next block of glass.

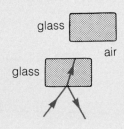

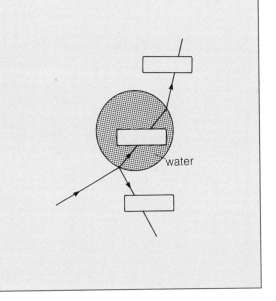

Refraction by a prism

EXPERIMENT. A prism deviates light

Lay a prism and a rectangular block of glass side by side. Send a ray of light through them both and look for two differences in the way that they refract light:

1. The ray that leaves the block is parallel to the ray going in, but the ray leaving the prism is not. The prism refracts the ray so that it comes out in a different direction. This is called **deviation**.

2. The block makes no difference to the colour of the light, but the ray leaving the prism has coloured edges. The prism produces colours from the white light. This is called **dispersion**.

Deviation and dispersion occur because the sides of the prism are not parallel and meet at an angle.

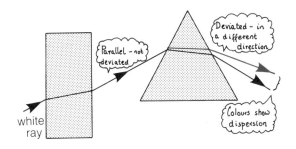

10 Copy this picture and draw in the normals where the ray strikes the surfaces. Use it to show that the rays are refracted at each surface in the same way as a block of glass.

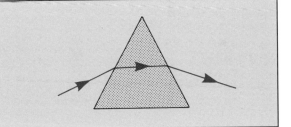

Colour

A prism can produce colours from white light. You can see these colours if you look at a window or lamp through the corner of a prism. The following experiments examine the colours more closely.

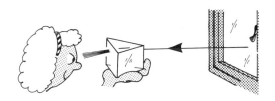

EXPERIMENT. Getting colours from white light

Use a ray box (or projector) with a slit to shine a ray of light through a prism. Arrange a white screen to collect the light after it has passed through the prism. Turn the prism until you get a rainbow of colours on the screen.

White light seems to be made up of coloured lights and can be split up into its colours by a prism. The prism separates the colours into a spectrum of the white light.

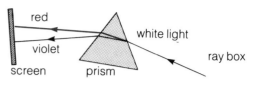
A prism can split white light into a spectrum

EXPERIMENT. Making a brighter and purer spectrum

Place a lens between the ray box and the prism of the experiment above. (It is best to remove the slit and use a wide beam of light.) Move the lens and turn the prism until the colours on the screen are as bright and sharp as possible. All the colours of the spectrum should be there with no white light in between. Check you can see the following colours: red, orange, yellow, green, blue, violet.

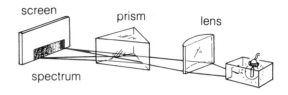
Making a brighter and purer spectrum

> **11** 'Rely on your gruff bass voice' is a sentence where the first letters of the words stand for the colours of the spectrum. It may help you to remember those colours. Can you think of a better sentence than this?

Colour separation

A glass prism can separate colours because the glass refracts each colour by a different amount. Two colours that have travelled together as far as the surface will be refracted through different angles and set off in different directions through the glass. This separation of colours from one another by refraction is the cause of dispersion. The colours are further separated by refraction when they leave the prism.

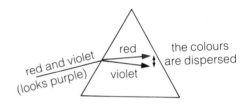

> **12** Copy and put the words **refraction**, **reflection**, **dispersion** and **deviation** in the correct boxes.
>
> | | |
> |---|---|
> | When light bounces off a surface | |
> | When light changes its speed and direction at a surface. | |
> | When light is made to travel in a different direction | |
> | When light is split into its separate colours. | |

A spectrum of beads

Separating the colours of white light into a spectrum is rather like sorting a mixture of small coloured beads into separate colours. A mixture of very tiny red, green and blue beads would not show a definite colour. But if the beads of each colour were put into separate piles, the colours would show strongly. The three piles of beads would be a colour spectrum of the mixture.

Other spectra

We often separate other mixtures into their different parts. The radio waves that surround us are a mixture of different stations. We use a radio set to separate the stations from one another and make a spectrum of the waves.

13 The pictures show other mixtures being separated into their spectra. Copy and in each box put the device that produces the spectrum. Use the following: radio set, TV set, prism, race, ear and brain.

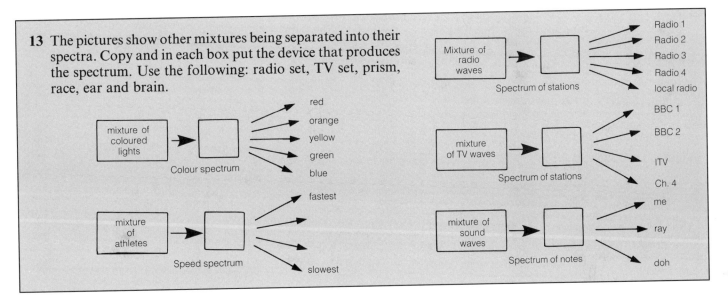

Critical angle

EXPERIMENT

Pass a ray of light into a semicircular block as shown in the pictures. Move the ray round as shown and notice that with small angles the ray is refracted and leaves the block. With large angles the ray is reflected from the inside surface of the block. And at a special angle in between, the ray just emerges along the surface of the block.

This special angle (between the ray and the normal inside the glass) is called the **critical angle** of the glass. It is about 42° for glass. If the ray hits the inside surface at an angle greater than the critical angle, all of the light is reflected. This is called **total internal reflection**. The surface acts as a perfect mirror.

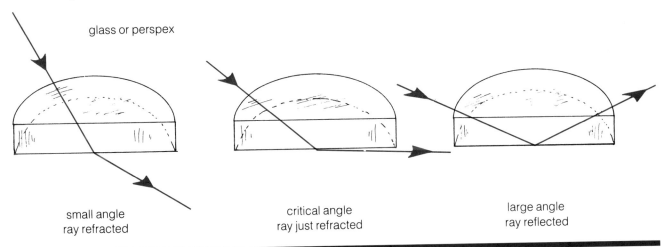

EXPERIMENT. Measuring the critical angle of Perspex

Place a semicircular block of Perspex on a sheet of paper. Aim a ray of light through the block at the centre of the straight side. Then turn the block until the ray just flicks over from being refracted to being reflected internally. Draw the direction of the ray by putting dots along it and draw round the block. You can then take away the block and construct the critical angle as shown. Use a protractor to measure this angle.

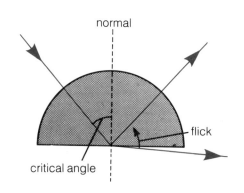

14 Copy and complete the picture showing what happens to the three rays of light. Label the rays:
1. refracted and reflected ray,
2. critical ray just refracted,
3. totally internally reflected ray.

15 The diagram shows rays of light from an underwater torch hitting the surface.
(a) Explain why some of the rays cannot get out of the surface.
(b) What would you see from above the surface of the water?

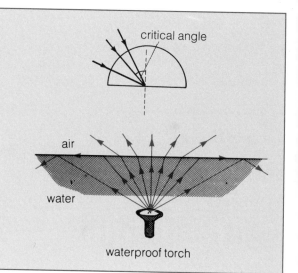

Total internal reflection by a prism

EXPERIMENT. Reflection by prisms

A right angled prism with base angles of 45° can be used to reflect light through exactly 90° or 180°.

Shine light rays into prisms like this and see if you can reflect the rays in the ways shown.

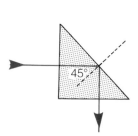
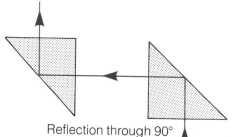

Reflection through 90°

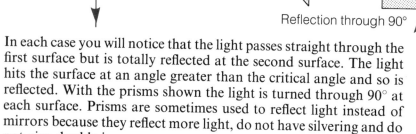

Reflection through 180°

In each case you will notice that the light passes straight through the first surface but is totally reflected at the second surface. The light hits the surface at an angle greater than the critical angle and so is reflected. With the prisms shown the light is turned through 90° at each surface. Prisms are sometimes used to reflect light instead of mirrors because they reflect more light, do not have silvering and do not give double images.

16 Is the critical angle for the glass of the prism on p. 130 more or less than 45°?

17 Draw diagrams to show a prism reflecting light rays in the directions shown.

18 Use a prism to read a book from behind and from the side. In each case say whether the words are upside down or upright; reversed or the right way round.

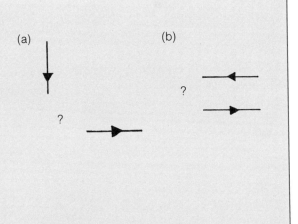

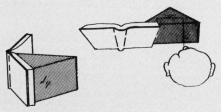

Uses of totally reflecting prisms

Periscopes
Useful for looking over peoples heads, looking round corners and for submariners!

Two prisms or mirrors reflecting rays of light in a periscope

19 Copy this diagram and draw the path of the ray from the top of the arrow through the periscope. Would you expect a periscope to give an upright image?

20 Draw the same two prisms in the positions of the ?'s so that the man can see light coming from behind.

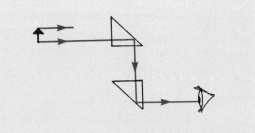

21 Red plastic prisms are used to make bicycle reflectors. Use the diagram to explain how they work.

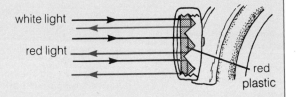

22 Draw a diagram of a periscope being used upside down.

Light pipes

Tubes of clear material such as glass and plastic can be used to 'pipe' light from place to place. The light becomes trapped in the tube by total internal reflection and travels along it with very little loss of energy.

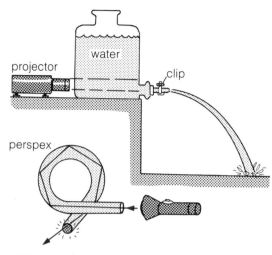

EXPERIMENT. Sending light by tube

The pictures show two ways of making your own light pipe. The first pipe is a smooth stream of water that flows from the bottom of a jar. Adjust the stream until the water flows without mixing. Shine light into its upper end and darken the room. You will see the light travel along the curved jet until it splashes into the sink. This liquid light pipe is flexible and as it moves the light moves too.

The second pipe is made from a Perspex rod that has been carefully heated and bent. Most of the light sent in one end finds its way around the bends to the other end.

When light enters these tubes, it travels in a straight line until it strikes the inside surface. This is usually at an angle greater than the critical angle of the material and all the light energy is reflected. The light can therefore zigzag its way along the pipe until it emerges from the far end.

Optical fibres

Optical fibres are light pipes made from thin glass or plastic threads. The fibres are flexible and can carry light to places that are too small or hidden to be lit with lamps. Bundles of fibres can be used to bring back images from places that cannot be seen directly. Each fibre in the bundle carries light from part of the object at its far end. For an image to be built up correctly, the fibres must be in the same order at each end.

A bundle can have two parts, one to carry light to a dark corner and another to bring back an image. The bundle can be fed into the human body for example and bring out pictures from deep inside.

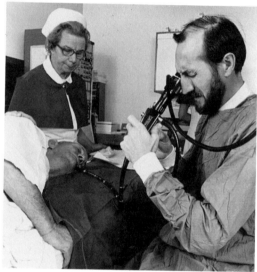

Using a light pipe to look inside a patient's stomach

23 The drawing shows a 'coherent' bundle of optical fibres. The fibres are arranged so that they are in the same order at each end. Copy the picture and number the fibres at the unnumbered end. If the bundle carries light from an object, would the picture at the far end be the right way round?

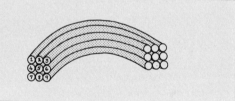

Lenses and rays

EXPERIMENT. A lens can focus light

Pass parallel rays of light from a ray box through a convex lens. You will see that the rays are refracted through a single point called the **focus** of the lens. The distance from the focus to the lens is called the **focal length**. Draw round the lens, mark the focus and measure the focal length. Repeat with a stronger lens (one that is more curved). You will find that a strong lens has a short focal length and a weak lens has a long focal length.

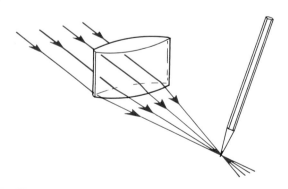

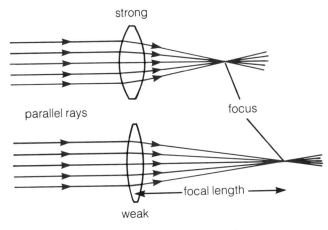

The principal focus of a convex lens

EXPERIMENT. The focal length of a 'round' convex lens.
The sunshine method

The rays of light from a point on the sun (or a distant window) are very nearly parallel. If they pass through a lens they will form part of an image at its focus.

Use the lens to form a small real image of the sun (or distant window) on a screen. This image is at the focus of the lens so measure from the screen to the centre of the lens to get the focal length.

EXPERIMENT. Measuring the focal length of a concave lens

Concave lenses are thinner in the centre than they are at the edges and make parallel rays of light spread out (diverge). To find the focus of a concave lens you must mark the diverging rays with dots and then draw the rays back until they meet. This type of lens has a negative focal length. Measure the focal length of a concave lens. Then put a concave and convex lens together and measure the focal length of the combined lens.

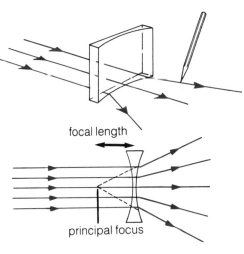

The principal focus of a concave lens

24 Choose a focal length for each lens from the values given. Copy and complete the table.

Lens	◖	◖	) (	)(
Focal length				
Strong or weak				

Choose from 20 cm; 5 cm; −5 cm; −15 cm.

How are images formed?

Real images

If a multiple slit is placed in front of a ray box, several rays of light will spread out from the lamp. If these rays are reflected by a concave mirror or pass through a lens, they can form into a cone of rays that travel towards a point. A real image is formed when light rays travel and meet like this.

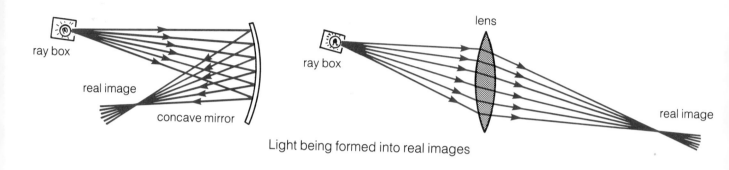

Light being formed into real images

Virtual images

Sometimes light rays leave a mirror, lens or water surface as a cone, but are moving away from a point. Although the rays never meet, they can be received by the eye and appear to be coming from a point. The eye 'sees' an image at that point although nothing is actually there. This is a virtual image.

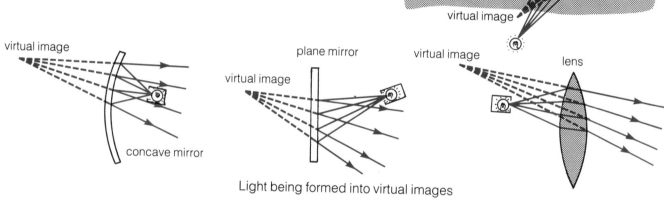

Light being formed into virtual images

25 Copy and carry the rays on through the lenses in these diagrams.

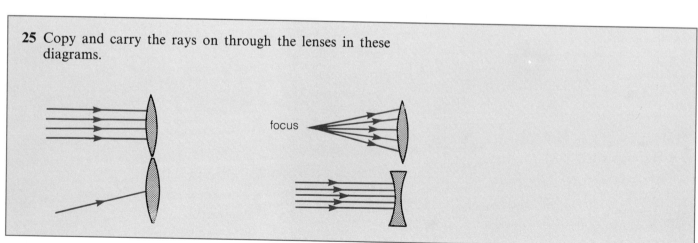

26 (a) Which of these lenses would you choose as a magnifying glass?
(b) Which lens would have the longest positive focal length?
(c) Which lens would have the shortest negative focal length?

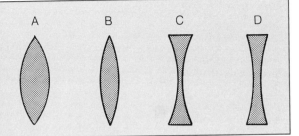

Lenses are used to form images in a number of important instruments.

The lens camera

A camera is a light-proof box with a lens at one end and photographic film at the other.

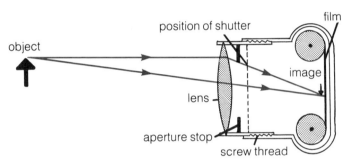

A simple camera

Focusing

The lens forms a small, real, upside-down image on the film (p. 119). This image must be focused by moving the lens 'out' for a close object and 'in' for an object in the distance. Marks on the lens help to set the lens for the correct distance.

Light control

It takes very little light to activate photographic film and capture an image. Too much light will spoil the film so it is important to control the amount of light that falls on the film. There are two ways of doing this.

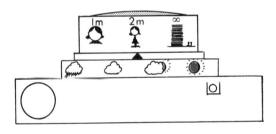

A camera has marks on the lens ring to help with focusing

Shutter opening times	1/30 s	1/60 s	1/125 s	1/250 s
	'long'			'short'

The shutter time control

The shutter stops light from reaching the film. The shutter can be opened for short times (e.g. 1/250 s) or longer times (e.g. 1/30 s). More light will fall on the film with long opening times than with short ones. Note that moving objects will appear blurred unless the opening times are very short.

27 Which of these words correctly describes the image formed on the film of a camera: upright, reversed, upside down, coloured, small, black and white, the right way round, large?

28 Copy and complete table.

Shutter opening times	1/30 s	1/250 s
Is this a long or short time?		
How much light will fall on the film, a lot or a little?		
Will the image of a moving cyclist be sharp or blurred?		

The aperture control

The amount of light that reaches the film can also be controlled by the size of a hole or **aperture** placed behind the lens. A large hole will allow a bright image to form on the film. The aperture control is marked in f-numbers or pictures. (An aperture with a value of f/8 means the diameter of the hole is $\frac{1}{8}$ of the focal length of the lens.)

The apertures needed for different lighting conditions

29 Copy and complete this table about the different parts of a camera.

Part	What it does
Lens	
Shutter	
Aperture stop	
Film	
Screw thread	

30 Which of the combinations of shutter times and apertures would you choose to give:
(a) the largest amount of light on the film,
(b) the smallest amount of light on the film?

	1/30 s	1/250 s
f/11	A	B
f/4	C	D

The slide projector

Properly designed projectors give a much brighter and clearer image than the home-made projector on p. 120.

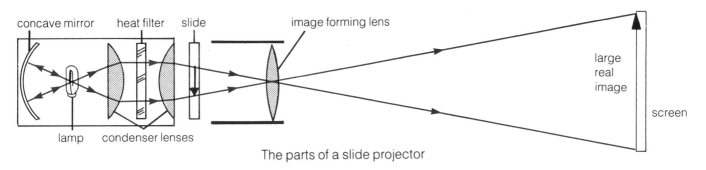

The parts of a slide projector

To get as much light as possible onto the slide, to produce a bright image, the designer has:
(a) used a very powerful (500 W) lamp,
(b) put a curved mirror behind the lamp to reflect the light forward,
(c) used two cheap lenses (called condenser lenses) to bend the light onto the slide.

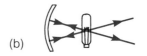
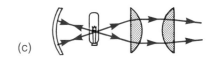

To remove heat and prevent it reaching and buckling the slide, the designer has:
(d) built in a cooling fan and put slots in the lid,
(e) used a piece of special glass, called a heat filter, that lets light through but stops infrared radiation (p. 100). (This radiation carries much of the heat away from the lamp.)

Finally a good quality lens is used to form an image of the brightly lit slide on a screen.

---

31 How must a slide be put into a projector to give an upright image, the correct way round?

32 What two things must be done to get a larger image?

33 What difference, other than size, is there between a large and small image?

34 What would happen if the heat filter were removed?

35 What would you do to the slide to correct these images of the letter R? Say whether to turn it upside down, to turn it round sideways, or both.

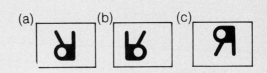

Back projection

The image can be much brighter if it shines onto a tissue screen and is viewed from the other side to the projector. Try it and see. Why do you think a brighter image can be seen this way?

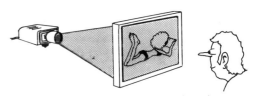

'Back projection' gives a brighter image

Light and Images check list

After studying this section on Light and Images you should:

- know that light is a form of energy
- know how to use a lens to form a real image and how to change the size of that image
- know how to use a convex lens as a magnifying glass to form an upright magnified image
- know the difference between a real and virtual image
- be able to describe the image made by a plane mirror
- know that light rays travel in straight lines
- know the laws of reflection
- understand how we see by scattered light
- know how to use rays to find an image in a plane mirror
- know how to use a concave mirror to focus rays to a point
- know how to use a concave mirror to send light out as a parallel beam
- understand what is meant by refraction
- know the path that a ray takes as it passes through a rectangular block
- know what happens to the energy of that ray
- know how a prism refracts light
- understand what is meant by deviation
- understand what is meant by dispersion and why it happens
- know how to split white light into the colours of the rainbow
- know how to use a lens and prism to make a pure spectrum of white light
- understand what is meant by critical angle and total internal reflection
- know how to measure the critical angle for glass
- know how to use 45° prisms to reflect light
- know some uses of total internal reflection
- know how to find the principal focus of a positive lens
- know the parts of a simple camera
- be able to describe the image it forms on the film
- know how to control the amount of light that falls on the film
- know the structure and use of a projector, the type of image it forms and how to adjust that image

Revision quiz

Use these questions to help you revise the sections on Light and Images.

- Name three things that use light energy.
  ... green plants (photosynthesis), animal eyes and photographic film or paper

- How can the image of a lamp, formed by a lens, be made larger?
  ... by moving the screen away from the lamp and moving the lens towards the lamp, to refocus the image.

- How can you use a positive lens to make an upright magnified image?
  ... by holding the object close to the lens and the lens close to the eye. This forms a magnified virtual image.

- Give four points about the image formed by a plane mirror.
  ... it is upright, reversed, virtual and an equal distance behind the mirror.

- What is a real image?
  ... it is an image through which light rays pass and which can be shone on a screen.

- What is a virtual image?
  ... it is an image which is formed where rays appear to come from and which cannot be formed on a screen.

- What is meant by a normal?
  ... it is the perpendicular at the place where a ray of light strikes a surface.

- Name the angle between the normal and (a) the in-going ray, and (b) the out-going ray.
  (a) angle of incidence, (b) angle of reflection or refraction.

- State a law of reflection
  ... the angle of incidence = the angle of reflection.

- Name two devices that use a concave mirror to reflect waves to a focus.
  ... a microwave dish aerial, a reflecting telescope.

- Name three devices that use a concave mirror to send out a parallel beam of waves.
  ... a torch reflector, a heater reflector and a microwave sending aerial.

- What is meant by refraction?
  ... it is when light changes speed and direction when it passes from one material to another.

- Which way is light 'bent' by refraction?
  ... towards the normal when it goes into a denser material (slows down). Away from the normal when it goes into a less dense material (speeds up).

- What happens to the light energy when it strikes a block of glass?
  ... some is reflected, the rest is refracted and enters the glass. Inside the glass some energy is absorbed and reflected, the rest gets out and is transmitted.

- Why is a prism said to deviate a ray of light?
  ... because the prism refracts the ray twice so that it comes out travelling in a different direction.

- What is the dispersion of light?
  ... it is when light is separated into its colours, e.g. by a prism.

- What is a colour spectrum?
  ... it is what you get when the colours in a mixed beam of light are separated out.

- What is meant by the critical angle of a material?
  ... it is the angle of incidence inside the material for which a light ray can just escape along the surface.

- When is a light ray totally internally reflected?
  ... when it strikes the inside surface of a material at an angle greater than its critical angle.

- Name three practical uses of total internal reflection.
  ... in periscopes, bicycle reflectors, optical fibres.

Examination questions

1 (a) Fig. 1 shows two rays of light leaving an object O and striking a plane mirror.
Draw the two reflected rays and use them to find the position of the image.
(b) Fig. 2 shows a side view of an electric fire.

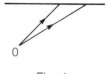

Fig. 1

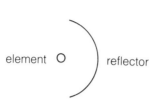

Fig. 2

   (i) What types of electromagnetic waves are given out by the element?
  (ii) What name is given to the shape of the reflector?
 (iii) The reflector is made of metal. Describe its surface, and explain why metal is used.

(NEA)

2 (a) Fig. 3 shows a long block of glass over an object O. Light from O reaches the top surface of the glass at X, Y and Z.
Copy and complete the following sentences.
  (i) The name given to the bending of the light at X is . . . . . .
 (ii) Fill in the two missing words in the following sentence.
     At Z light is . . . . reflected.
 (iii) The angle marked R is called . . . . . .
 (iv) Light is reflected as shown at Z because . . . . . .
(b) Fig. 4 shows two 45° 45° 90° glass prisms with two rays of light incident on a face of one of them.
  (i) Copy and complete the path of both rays through both prisms.
 (ii) A practical use for such a device would be . . . . . .

(NEA)

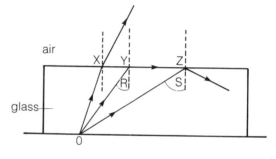

Fig. 3

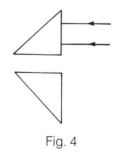

Fig. 4

3 (a) The diagram below shows a ray of red light incidence on a glass prism.
Copy and complete the diagram to show the path of the light through and out of the prism.
(b) If white light were used instead of red light on this prism, what difference, if any, would you notice?

(MEG)

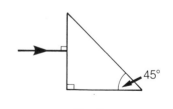

Fig. 5

15 Heat at work

Heat is a form of energy that passes from hot to cold objects. In the following experiments, heat passes from a hot flame into a colder substance. The substance warms up and a number of things are seen to happen. Set up the experiments if you can and watch them carefully.

EXPERIMENT. Looking at what heat can do

1. Stretch a strip of aluminium foil between two jars and heat it with a row of candles.

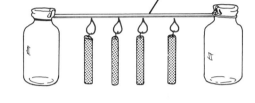

2. Twist the ends of pieces of copper and iron wire together. Connect the other ends to a sensitive current meter and heat the twisted joint with a bunsen flame.

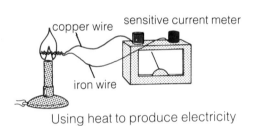

Using heat to produce electricity

3. Light the candle under one of the chimneys in an apparatus like this. Use smoke from burning cardboard to show up any movement of the air caused by heat from the flame.

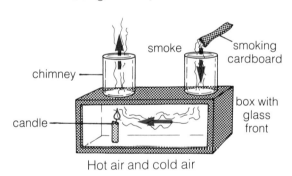

Hot air and cold air

4. Make a simple turbine by cutting a spiral from a circle of card. Tie a thread to the centre of the spiral and let it hang over a bunsen flame.

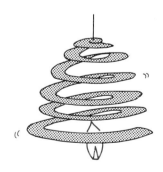

1 These four experiments show that:

Heat energy can be changed into kinetic energy.
Heat makes metals expand by a small amount and get longer.
Heat can be changed directly into electrical energy.
Heat can produce hot and cold currents of air.

Which experiment leads to each of these conclusions?

Hot gas engines

We would not get far without 'heat engines'. These are engines that use fuel to make hot gases that expand and produce movement. They have helped to put people on the moon, they power cars and boats, fly aeroplanes across continents and drive generators that make the electricity we use.

2 Five heat engines in common use are: rockets, petrol engines, diesel engines, jet engines and steam turbines. Copy this table of machines and say which type of heat engine each could use.

	Heat engine
Cars	
Lorries and boats	
Electric power stations	
Aeroplanes	
Satellite launch vehicles	

Steam turbines

Coal and nuclear electric power stations use steam turbines to change the heat they produce into electricity. At the heart of the turbine is a wheel that has a set of blades fixed to a revolving shaft. Water is boiled under great pressure, producing very hot, high pressure steam. This steam (the hot gas that the engine uses) is made to hit the blades of the turbine wheel. As the steam shoots off the curved blades, it pushes the wheel round. A carefully designed turbine can use about a third of the energy of the steam to keep its blades turning.

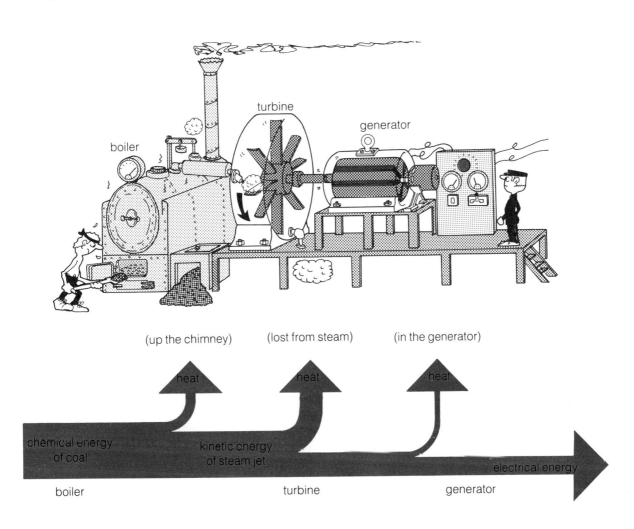

---

3 Look at the energy changes that happen in a coal-fired power station. Describe what happens to the energy as it is changed by the machinery.

4 Does most energy escape in the boiler, the turbine or the generator?
Give a reason why this might be.

Jet and rocket engines

The small boy in the picture is trying to get moving by pushing on the big boy. The same force that moves him to the right acts on the big boy and moves him to the left (see Newton's third law, p. 4). The forces give the boys equal amounts of momentum in opposite directions, but because their masses are different their speeds will be different. Similarly if you blow up a balloon and let go of the neck it will fly about the room. The air rushes out in one direction and pushes against the balloon, making it fly in the other direction.

Jet and rocket engines work in a similar way. They produce very hot gas in a container and let it out through a hole at the rear. As the hot gases rush out backwards, the container is pushed forwards.

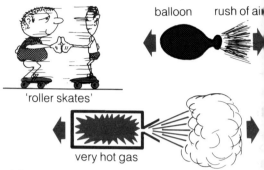

High pressure gases producing movement

5 Copy this table and use it to compare the small boy with the large boy and the balloon with the air. Put 'small', 'large' or 'same' in the boxes.

	Small boy	Big boy	Balloon	Air
The size of the force on them				
Masses				
The amount of momentum gained				
Speed				

Jet engines (gas turbines)

These engines can develop tremendous power and are especially useful for aircraft. They take air, compress it and squirt in paraffin. The paraffin and air burn, producing very hot gas. This gas rushes out of the jet nozzle and the engine is forced forwards.

Some of the energy of the hot gas is used to keep turbine blades turning that compress the air as it enters the engine.

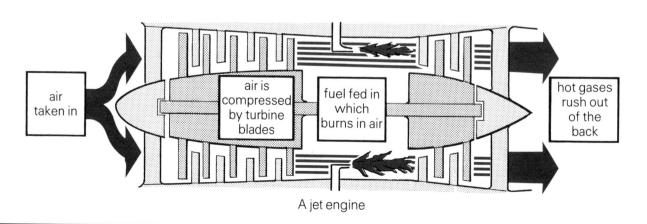

A jet engine

6 Describe the energy changes that take place as a jet engine is started up.

7 Why are jet engines not used in space or on the moon?

Rocket engines

Rocket engines carry fuel and air (or rather oxygen). The fuel and oxygen burn producing hot gas that escapes and gives the forward movement. A rocket will work where there is no air. Since most of the mass of a rocket is fuel, it gets much lighter as it flies.

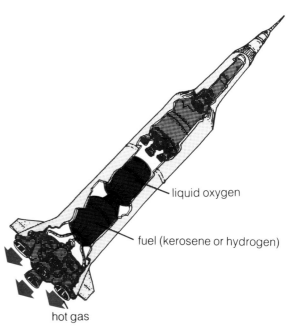

A rocket engine

8 A satellite launch vehicle has a three-stage engine. The first-stage rocket burns until it runs out of fuel and then drops off. The second stage then takes over and burns until it runs out of fuel and drops off. The third stage then ignites and accelerates the satellite until its fuel is finished.

Explain why a three-stage engine like this can give a satellite more speed than a single-stage engine with the same mass of fuel.

9 Is a firework 'rocket' a jet engine or a rocket engine? Where does the hot gas come from that makes it fly?

Petrol engines

Petrol engines produce hot gas from an explosion. The gas does not rush out but is made to move a piston instead. Small air-cooled petrol engines come in many shapes and sizes but they all have the following main parts:

Petrol tank and tap
Air intake and filter
Carburettor
Throttle cable
Cylinder block and crankcase
Spark plug
High voltage lead
Cooling fins
Exhaust and silencer

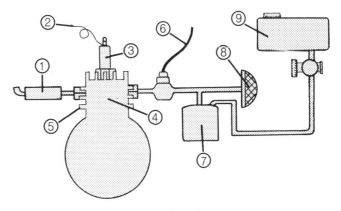

A petrol engine

10 Draw this diagram of a single-cylinder petrol engine. Write down the list of parts and try to number them correctly.

145

The fuel used in this sort of engine is a mixture of petrol droplets and air. This mixture is exploded in the cylinder by a spark, producing a very hot gas. The gas expands pushing the piston down.

This table shows what some of the parts do.

Part	The job it does
Air filter	Removes dirt from the air sucked into the engine.
Carburettor	Mixes small drops of petrol with the air.
Throttle	Controls how much air and petrol enter the engine and so the speed of the engine.
Spark plug	Sparks at the right moment and explodes the petrol/air mixture.

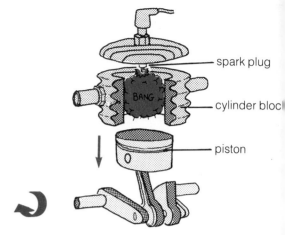

Rotation from up and down motion

11 Would a petrol engine work on the moon? Explain.

12 Copy this energy-flow diagram for a petrol engine starting up. Write in these forms of energy instead of the question marks: kinetic energy; chemical energy; heat energy.

13 What happens to a petrol engine if you place your hand over the air intake when it is running? Explain why.

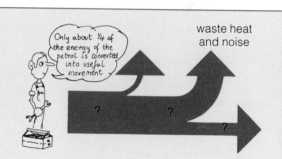

What happens inside a common type of petrol engine?

The piston of a 'four-stroke' engine goes down, up, down, up, each stroke having a different effect on the gases in the cylinder. The four stokes are then repeated, always in the same order.

1. Intake stroke
A 'four-stroke' engine has two valves. As the piston moves down the pressure is reduced in the cylinder. The inlet valve then opens and petrol/air mixture is forced into the cylinder.

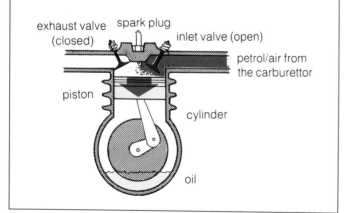

2. Compression stroke
Both valves close and the petrol/air mixture is compressed to about one tenth of its starting volume.

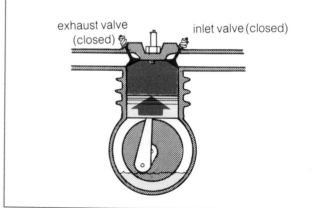

3. Power stroke

The spark plug sparks just as the piston is starting to move down. The petrol and air explode, producing a very hot gas. This gas expands and forces the piston down.

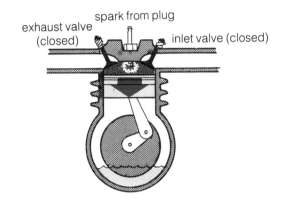

4. Exhaust stroke

The movement continues pushing the piston up. The hot gas is driven out through the open exhaust valve.

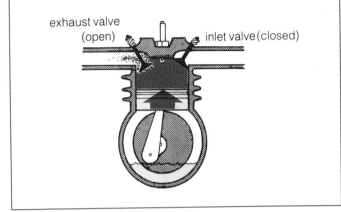

14 Copy and complete.

Stroke	Inlet valve	Exhaust valve	What is happening to the fuel	Which way the piston is moving
1. Intake (suck)				
2. Compression (squeeze)				
3. Power (bang)				
4. Exhaust (blow)				

Open or closed?   Up or down?

15 Which stroke provides the useful energy?

16 Copy and fill in this table about heat engines.

	Does it take in air?	Does the hot gas make machinery go round?	Does the escaping gas drive the engine forwards?	What uses this type of heat engine?
Steam turbine				
Jet engine				
Rocket engine				
Petrol engine				

16 The expansion of solids

⊙ **EXPERIMENT. Heat causes expansion**

Build this piece of apparatus and watch the pointer as the bar is heated. Notice how the roller and pointer show up the tiny expansion (growth) of the bar. Which way does the pointer move?

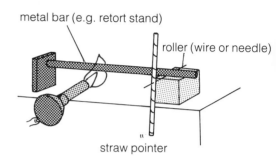

1 (a) How could you change the apparatus to magnify the expansion even more (give two ways)?
  (b) If the circumference of the roller is 1 mm and the pointer moves through 45° how much did the bar expand?

EXPERIMENT. Measuring the expansion of metals

In this apparatus, steam is used to heat the metal. The levers magnify the expansion about 50 times, making it easy to see and measure. Measure the expansion of different metals using this or similar apparatus.

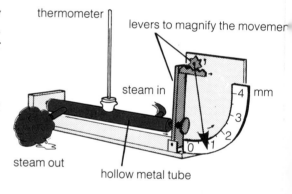

Results. Tubes or bars of metal 1000 mm long and heated from 0°C to 100°C would expand by the following amounts:

Copper 1.7 mm
Steel 1.2 mm
Aluminium 2.3 mm
Other materials expand too:
Glass 0.9 mm
'Pyrex' glass 0.05 mm
Concrete 1.2 mm

The expansion is small. It can cause problems and it can be useful.

2 When steel is heated we have seen that it increases in length.
  (a) Would you expect a steel ruler to increase in width and thickness too?
  (b) Would a hole in the ruler get bigger or smaller?
  (c) If you measured a length using a hot steel ruler, would the value you get be too long or too short? Explain your answer.

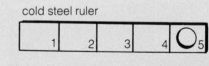

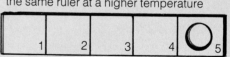

The increase in size is exaggerated about 100 times

The force of expansion and contraction

This apparatus can be used to show that large forces are exerted when materials expand or contract. The metal rod is headed and as it expands, the screw is tightened to take up the slack. The bunsen is then removed and the rod allowed to cool. As it attempts to contract, an increasing force is exerted on the iron pin until it eventually breaks.

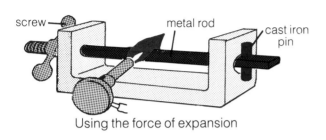

Using the force of expansion

Problems caused by expansion

3 What do you think will happen to the loop in the metal tube when it is heated. What will happen to the pipe without a loop? Their ends are firmly fixed.

(Steam and oil pipelines in hot countries sometimes have loops in them to allow for expansion.)

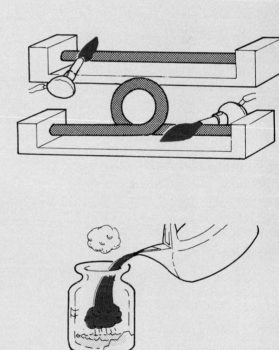

4 If you pour boiling water into a thick glass jar, it will probably crack. The inside expands, the outside does not and the stress cracks the glass. A jar made from Pyrex glass however would not crack. Look at the expansion figures and explain why not.

5 Why are gaps left between the slabs of a concrete road? Why is tar such a good material to use to fill these gaps? What would happen if no gaps were left? Should gaps be left at the edge of the road?

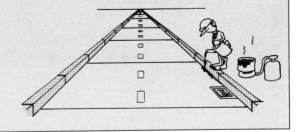

6 A steel cap, screwed tight on a glass bottle, can sometimes be loosened by pouring hot water over it. Look at the expansion figures for glass and steel and explain why this should work.

Why does heat make solids expand?

When a solid is heated its molecules vibrate with more energy and move further away from each other. The solid gets a little larger in every direction.

The 'expansion' of a line of people

The bimetal strip

A bimetal strip is made of two metals firmly welded together. Steel and copper are examples of two metals that could be used.

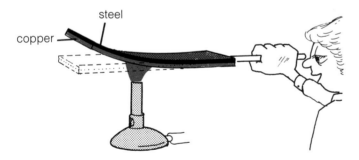

Bimetal strips bend when heated or cooled

7 If you heat a bimetal strip, it bends. Look up the expansion figures for steel and copper (p. 148) and explain why this happens.

8 If the strip is straight at room temperature and you plunge it into ice, what will happen?

EXPERIMENT. Thermostats – automatic temperature controllers

Obtain an electric iron with a light that shows when the iron is on. Switch on (the light will come on) and start the clock. After a while, when the iron is hot enough, the light will go off. Note the time when this happens. The iron will then continue to go on and off by itself. Write down the times when it goes on and when it goes off. The iron has a thermostat inside; a switch that goes on when the temperature is too low and off when the temperature is too high.

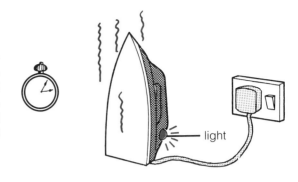

Investigating the action of a thermostat

9 An electric iron was found to go on and off at the following times:

On	Off	On	Off	On	Off	On	Off
0	3 min 15 s	3.30	4.00	4.10	4.20	5.00	5.05

Plot an on/off time diagram on graph paper. Calculate the time that the iron is on. Work this out as a percentage of the total time.

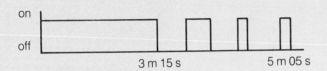

10 Write down two advantages of having a thermostat in an iron.

11 Why is the iron on so long at first compared with later.

How does the thermostat work?

The thermostat is made from a bimetal strip, a metal bar and two contact points. The current for the heating plate has to pass along the bimetal strip and between the contact points. When the plate is hot enough the bimetal strip bends and the current is switched off.

closed (cold) current flows

open (hot) no current

A bimetal thermostat

151

12 Continue this account by explaining what happens as the heating plate cools.

13 Copy and complete this table. It is about the action of a thermostat in an electric iron.

	Is the strip straight or bent?	Is the current on or off?	Are the points in contact?
Temperature high			
Temperature low			

14 **Bimetal thermometer.** This diagram shows a bimetal thermometer. When the temperature rises, the bimetal strip bends more (into a tighter coil) and the pointer moves round the scale.
   (a) Should the metal that expands the most be on the inside or the outside of the coil?
   (b) Write down two advantages that this thermometer has over a mercury-in-glass thermometer.

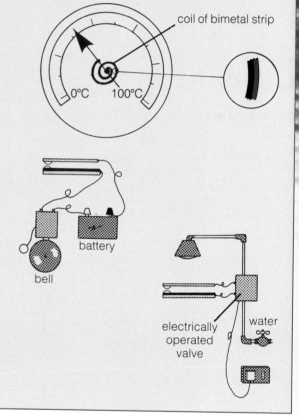

15 The picture shows a model fire alarm. Explain how it works.

16 The picture shows an automatic fire extinguisher that turns on a water spray when there is a fire. Explain how it works.

The expansion of liquids

You can see that liquids expand with this simple apparatus. The test-tube must be filled to the top so that when the bung is pushed in, liquid goes up the narrow tube. Warmth from a small flame should then cause enough expansion to be seen.

17 Describe how you could compare the expansion of different liquids. Name two things that you must do to make the comparison fair.

18 Liquid thermometers work by expansion. Which thermometer, A or B, will be the most sensitive? (i.e. show the most expansion for the same rise in temperature.) Explain your answer.

Convection currents

EXPERIMENT. Looking for convection currents

Put water into a flask, add a pinch of fine aluminium dust (care!) and two drops of washing-up liquid. The aluminium dust forms a 'snow storm' that shows up any movement of the water. Heat or cool the water by the different ways shown. Watch the moving currents of warm and cold water (convection currents). Draw simple diagrams of these convection currents. Which way do hot water currents move? Which way does the cold water move?

water and aluminium dust and washing-up liquid

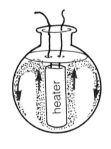

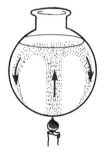

Looking at convection currents

Expansion and the convection of heat

'Convection' is the movement of heat energy by currents of hot liquid (or gas).

When a drop of liquid is heated it expands – its volume increases. The amount of material (its mass) does not change. Since its mass is more spread out, hot liquid is less dense than cold liquid around it. So in a mixture of hot and cold liquid, the cold liquid will sink to the bottom and the hot liquid will rise to the top. This is why convection currents flow.

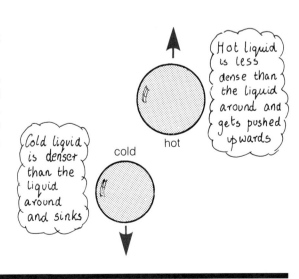

19 Copy these diagrams and put an arrow where you would have to heat the water to get the convection currents shown.

 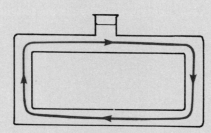 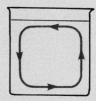

20 (a) Explain how the water at the top of this kettle gets heated by a heater that is at the bottom.

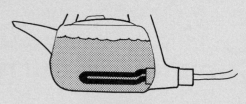

(b) Explain why the water in this tank that is below the immersion heater cannot be heated by convection currents of hot water. (See p. 160 also.)

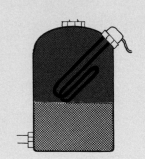

A hot and cold water system

Nearly all homes have 'running' hot and cold water. Convection plays a useful part in a household hot and cold water system. The one shown here has an electric heater in the hot water storage tank to heat the water. There are some special points to remember in designing such a system.

Special design feature	Reason
The cold water from the storage tank must enter the bottom of the hot tank.	The cold water falls to the bottom so with this design the hot and cold water do not mix. The hot water stays on top.
The hot water must be piped from the top of the hot tank.	The 'lighter' hot water will be at the top of the tank.
There must be a 'vent' pipe.	This allows steam or air bubbles to escape from the hot tank.

21 Copy this diagram and list of parts. Put the right number against each part.
   Cold water storage tank
   Hot water tank
   Heater and thermostat
   Hot tap
   Cold tap
   Drinking water tap
   Ballcock
   Rising main
   Vent pipe
   Overflow pipe.

22 Why is the drinking water tap, usually the one in the kitchen, connected to the rising main?

23 Draw in one stop tap that can shut off all the water. Draw in a second stop tap that can shut off just the hot water.

24 Why is the (hot) water level in the vent pipe slightly higher than the water level in the cold storage tank?

25 What is the job of the thermostat and what would happen if it stopped working?

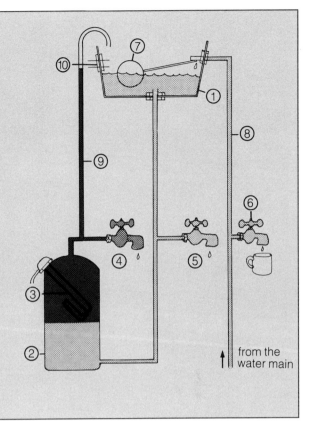

The expansion of air

The pictures show simple experiments you can do to watch the expansion of air when it is heated (and its contraction when it cools).

The air in the balloon expands when it is boiled in a saucepan and contracts when you take it out to cool.

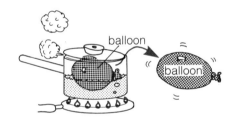

The experiment with air in a flask has three stages. The air is first warmed with the tube under water. Then the air is allowed to cool with the tube still under water. And finally the air is heated with the tube upright.

26 Explain carefully what you see at each stage and why it happens.

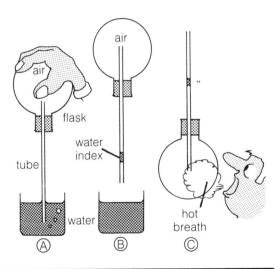

You will notice from this work that air expands much more than solids or liquids heated through the same temperature rises.

Convection of air

Air expands when it is heated and becomes less dense. So in a mixture of warm and cold air, the cold air will fall and force the warm air to rise. Here is a way of showing up these convection currents in air.

EXPERIMENT. Looking at convection currents in air

Arrange a projector so that it casts the shadow of a candle on a white screen. The shadow of the hot gases from the flame and the liquid wax should also be visible. Which way do the gases move? Experiment with this current of hot gas and air:

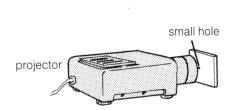
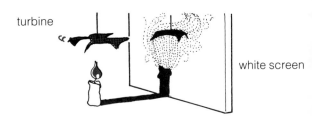

Showing up convection currents in air

(a) Place a turbine blade over the flame.
(b) Place a heat-resistant board over the flame to act like the ceiling of a room.
(c) Place a chimney over the flame.
In each case notice what happens to the rising warm air.

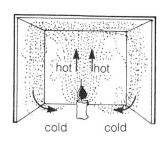

27 It is clear that warm air rises and so must be lighter than cold air. Can you explain

(a) how a flame can lift a man into the sky;

(b) why a kilt keeps a Scotsman's knees warm;

(c) why a fire or heater keeps air moving around a room;

(d) why the freezing compartment in a refrigerator is placed at the top;

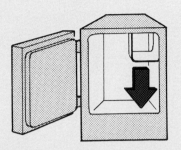

(e) why birds can sometimes fly for hours without flapping their wings;

(f) why it is better to crawl on the floor of a smoke-filled room.

Measuring the expansion of air

EXPERIMENT

The air for this experiment is trapped in a capillary tube by an index of sulphuric acid. The thread of acid dries the air and allows it to expand freely, without increasing its pressure. Fix the capillary tube and a thermometer to a millimetre scale. Put the closed end of the tube at the zero of the scale and the thermometer bulb close to the trapped air. Immerse the whole assembly in water and measure the length of trapped air at the temperature of the water. Use a bunsen to heat the water by about 10 °C, remove the flame and stir well. Measure the new length of the air column and the exact temperature of the water. Repeat this procedure until the water boils.

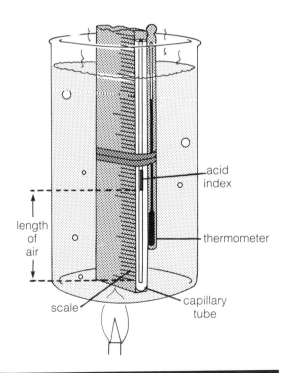

How to use your results to find the absolute zero of temperature (compare with p. 65)

When air cools it contracts. At a low enough temperature it would (in theory) contract to zero volume. This temperature would be the absolute zero of temperature. Get a rough value of 'absolute zero' by plotting your readings on the axes shown. Draw a straight line through the points and extend it to find the temperature at which the volume becomes zero.

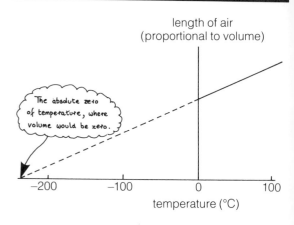

28 Use these readings of the length of a column of air and its temperature to get a value for the absolute zero of temperature.

Length (mm)	150	156	161	167	172	177	183	189	194	199	205
Temperature (°C)	0	10	20	30	40	50	60	70	80	90	100

How to see if your readings follow Charles' law

This states that:

The volume of a fixed mass of gas is proportional to its Kelvin temperature, provided the pressure of the gas is kept constant.

This means that

$$\frac{\text{volume of gas}}{\text{temperature in kelvin}} = \text{constant}$$

To see if your readings fit this law, convert each temperature to kelvin and divide it by the length of the air column (which represents the air's volume).

Your values should be constant enough to suggest that dry air does obey Charles' law.

29 These readings are of the volume of a gas and its temperature as it is heated from 0 °C to 100 °C.
(a) Convert the temperature reading to kelvin.
(b) Calculate temperature/volume for each temperature.
(c) Say whether the gas follows Charles' law.

Volume (cm³)	90	93	97	100	103	107	110	113	116	120	123
Temperature (°C)	0	10	20	30	40	50	60	70	80	90	100
Temperature (K)											
Kelvin temperature / volume											

17 The conduction of heat

EXPERIMENT

Hold a piece of copper wire in the flame of a burning match. Which do you have to drop first, the copper or the match? Heat travels quickly through the copper wire. It cannot be seen as it travels but it can be felt when it reaches your fingers. The movement of heat energy like this, without any obvious movement of the material, is called **conduction**.

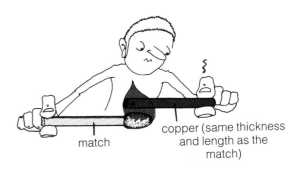
match / copper (same thickness and length as the match)

EXPERIMENT. How well do materials conduct heat?

Build this apparatus to see how well heat travels through a metal rod. Use grease to stick drawing pins at regular intervals along the rod. Heat one end with a bunsen and count the number of pins that fall off. This number is a rough measure of how well the rod conducts heat.

Use other similar rods made from different metals and find out which are the best conductors.

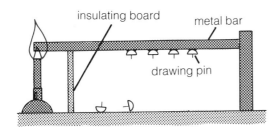

1. Which of our senses can tell us whether heat is being conducted through a material?

2. How would you use a match to show that wood does not conduct heat as well as a brass screw?

3. Should these things be good or bad conductors of heat? Copy and complete.

	Hot-water bottle	Saucepan	Blanket	Radiator	House bricks	Gloves
Good or Bad?						

4. Explain why the heated end of the iron rod in the experiment above glows red hot while the end of the copper rod does not.

EXPERIMENT. Is water a good or bad conductor?

Fill a long glass tube with cold water and hold it at a slant over a small flame. Heat the water gently in the middle of the tube and place a small amount of powdered dye in the water at the top. Watch the die as it shows up the movement of the water and feel the top and bottom of the tube to test the temperatures.

What can you conclude from the experiment?

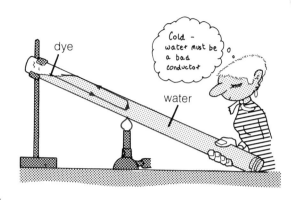

> **5** A student found that the dye only moved in the water above the flame and that the water at the top got very hot, while the water at the bottom stayed cold.
> Give a reason
> (a) why heat can travel easily in the water above the flame;
> (b) why very little heat travels through the water below the flame.

EXPERIMENT. Is air a good or bad conductor of heat?

Fit the thermometers into the ends of a cardboard tube and place a heater close to one end. Leave the apparatus for a while and watch the thermometers. When they become steady make a note of the readings.

Next turn the tube so that it is upright and put the heater at the bottom. Again wait and note the temperatures when they become steady.

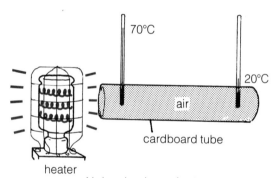

Air is a bad conductor

The temperature readings should show you that:
(a) very little heat travels along the tube when it is horizontal. Air must therefore be a bad conductor of heat.
(b) much more heat travels up the tube when it is vertical. This shows that air carries heat very well by convection if it is allowed to move upwards.

Good insulators like feathers, woolly jumpers and polystyrene contain many small pockets of air. This air is trapped and so cannot carry heat away by convection. The air pockets are bad conductors of heat and make very effective insulators.

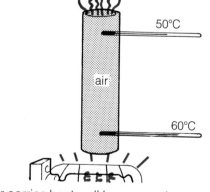

Air carries heat well by convection

EXPERIMENT. Testing insulators

An insulator is a material through which heat passes very slowly. Find out how well cotton wool and feathers stop heat escaping by using them to insulate cans full of boiling water. Have one can without any insulation as a control and find out how long each can takes to cool from 90 °C to 70 °C. Which material is the best insulator?

6 Copy the table and put 'good' or 'bad' in the boxes to show whether copper, water and air are good or bad conductors and insulators.

	Copper	Water	Air
Conductor			
Insulator			

7 Why can no heat pass by convection or conduction through a vacuum?

8 Explain why the air trapped in cotton wool and feathers is good at stopping heat loss and why a hot can surrounded by air loses heat so quickly.

9 Explain why string vests (on people) and feathers on birds are such good insulators of heat. (Note: a layer of still air is a good insulator but air that can move carries heat away by convection.) Would a string vest be warm on its own?

10 Explain the terms conduction and convection.

18 The radiation of heat

EXPERIMENT

Place the back of your hand by the side of an electric heater. Very little heat reaches the hand by conduction or convection and yet the skin gets very warm. There is a third way that heat energy can travel and that is by electromagnetic waves. We have seen that these waves carry energy (see p. 100). Hot objects such as this heater and the sun give out electromagnetic waves. This radiation then passes through air and space, spreading energy around at the speed of light. When the radiation falls on things, some of it soaks in and warms them up. The hot sun warms the Earth by radiation in this way.

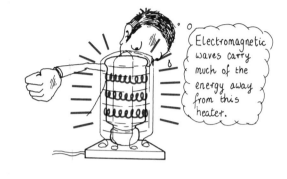

1 Why can we be sure that:
   (a) very little heat reaches the skin by conduction and convection;
   (b) heat does not reach the Earth from the sun by conduction and convection?

EXPERIMENT. Can radiation be reflected and focused?

Place the heater at the focus of a large concave mirror. Look straight into the mirror and you will feel and see the reflected radiation. The mirror reflects the waves from the heater into a nearly parallel beam that can be felt all over the face (see p. 126).

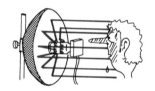

Use a second mirror to collect the beam and focus it into a hot spot. Find this spot by hand and then hold a live match there. The temperature should be sufficient to light a match. Radiation can be reflected and focused by mirrors.

heater at the focus of a concave reflector

match

Reflecting and focusing radiation

2 Which of the heaters shown:
   (a) focuses radiation;
   (b) reflects radiation into a broad beam?

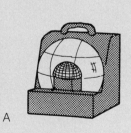  

Black is best

When radiation falls on a surface, some of its energy is absorbed and warms up that surface.

EXPERIMENT. Which surface is best at absorbing radiation?

Cover the back of your hand with a square of aluminium foil and hold it about 10 cm from a heater. Do the same experiment with foil that has been painted black. Can your skin tell you which surface is best at absorbing the radiation? If the radiation does not soak into the shiny surface, it must be reflected by it. If the black surface absorbs the radiation it cannot reflect it.

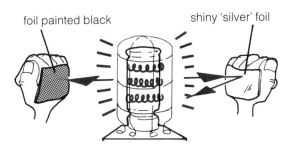

Black absorbs more than silver

3 Copy and write 'shiny' and 'black' in the correct box.

Good absorber		Bad reflector	
Bad absorber		Good reflector	

4 Which surface colour (black or shiny) would be best for:
   (a) a Rolls Royce in a hot country (no expense spared);
   (b) a space suit for the moon (very hot in the sun);
   (c) a solar water heater (used to collect the sun's heat).

5 On a sunny day, some pebbles on a beach get hotter than others. Explain why this might be.

EXPERIMENT. Which surface sends out the most radiation?

A sensitive radiation detector called a thermopile is needed for this experiment. Heat a shiny pot by filling it with hot water. Pick up the radiation from its surface and note the position of the light spot on the meter. Then paint the shiny surface black. The black surface is no hotter than the shiny surface, yet you will find it gives out much more radiation. A black surface is a good radiator as well as a good absorber.

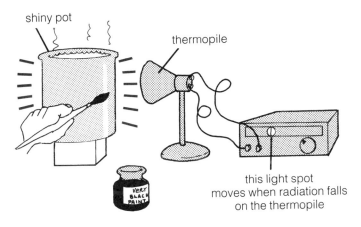

Black radiates more than silver

The vacuum flask

EXPERIMENT. How good is a vacuum flask at keeping things hot?

Put some hot water in a cup and an equal amount in a vacuum flask. Leave to cool and measure the temperatures after 30 minutes. Does the vacuum flask stop heat escaping from the water? Does it make any difference if you put a stopper in? Do an experiment to find out.

How does a vacuum flask work?

The flask has a double glass wall with a vacuum between the walls. The inside surfaces of the glass are silvered. The flask is designed to reduce the three ways that heat can escape.

Conduction

The vacuum stops all the conduction through the walls. A little heat is conducted up the walls and out that way.

Convection

The cork stops the air, heated by the water, from escaping with its energy. The cork will get warm and conduct a little heat out.

Radiation

This can cross the vacuum. The hot inside silver surface is a bad radiator so only a little radiation sets off across the vacuum. When it reaches the other side it is reflected back by the silvering on the outer wall. Very little radiation escapes.

Put an ice/salt mixture into a vacuum flask and you will find that it is just as good at stopping heat from getting in.

---

6 Why does the liquid in a vacuum flask not wash the silvering off the walls.

7 Liquids in vacuum flasks do not stay hot for ever. Explain three ways that heat can escape.

8 Explain why the two silver surfaces are good at stopping radiation from getting into the flask when it is holding cold liquid.

Keeping warm at home

Keeping a house full of heat energy is rather like keeping it full of water. Water would have to be pumped in all the time to replace the water that leaks out. If there are no leaks, no water has to be pumped in.

In cool climates heat also leaks out of houses, through roof, walls, floor, windows and doors. The heat that escapes has to be replaced to keep the house warm.

Less heat is lost if you insulate the house and stop draughts. It then takes less heat energy to replace the loss and keep the house warm.

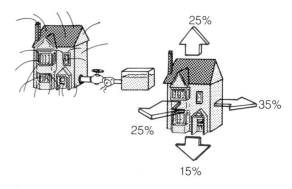

9 These lists show the main ways that a house loses heat and ways that this loss can be reduced. Rewrite the lists so that they match up.

Ways of losing heat
by conduction through the ceiling and roof
by conduction through the windows
by conduction through the floor
by conduction through the walls
by cold draughts and the escape of warm air.

Ways of reducing heat loss
(a) carefully seal doors and windows
(b) carpets on the floor
(c) cavity walls filled with foam
(d) two layers of glass instead of one
(e) thick insulation in the loft.

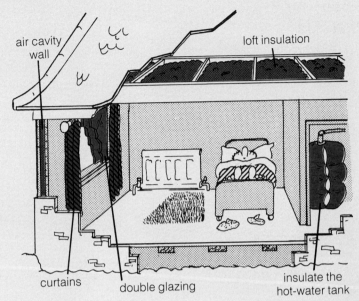

10 Find four other good insulators in this picture. Say what is helped to keep warm by each one.

U values

The amount of energy that is lost from a house can be calculated from the U values of its walls and roof. U values give a measure of the heat energy that passes through a wall by conduction. (The U value of a wall is the energy that passes each second through 1 square metre when one face is 1 °C hotter than the other.)

Good insulators have small U values and buildings in cold climates should be made from surfaces that have small U values. The pictures show how the U values of common building materials can be reduced by insulation and double glazing.

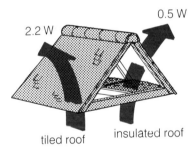

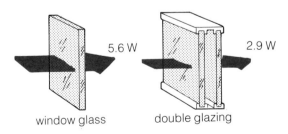

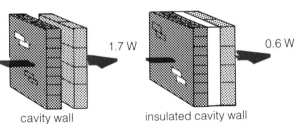

The heat lost each second through one square metre of wall for a 1°C temperature difference

U values depend on wind speed because a wind outside increases the amount of heat lost through a wall. The U values that follow are given in watts/square metre kelvin and are for still air.

Calculating the heat loss

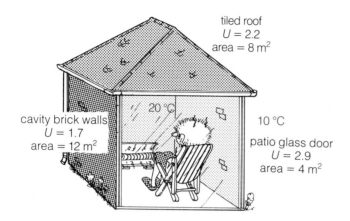

The heat lost through the walls and roof of this strange little house can be calculated using the equation:

heat lost/second = U value × area × temp. difference across the wall

    loss through the walls: $1.7 \times 12 \times 10 = 204$ watts
    loss through the patio door: $5.6 \times4 \times 10 = 224$ watts
    loss through the roof: $2.2 \times8 \times 10 = 176$ watts
                                 total loss = 604 watts

(The electric fire and the lady's body must produce heat at this rate to keep the house temperature at 20 °C.)

---

11 How much heating power would be needed to keep the house warm (at 20 °C) if the cavity walls were filled with foam, the roof insulated and the patio doors double glazed? (Use the figures given on p. 165.)

12 The pictures show other insulating materials in action. Use the values given to calculate the loss of heat/second for each example.

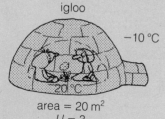

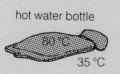
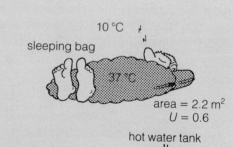

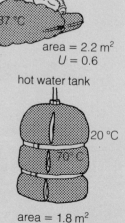

House radiators

Radiators full of hot water are often used to replace the heat lost from a house and so keep it warm.

Radiators use conduction, convection and radiation to transfer heat from the hot water inside to the air outside.

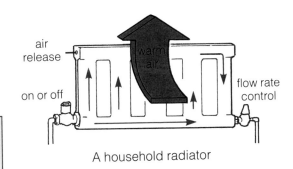

A household radiator

> 11 This table describes the ways that heat is transferred by a radiator to a room. Copy and say whether each method is conduction, convection or radiation.
>
> | | |
> |---|---|
> | Heat from very hot water passes through the metal radiator | |
> | Air in contact with the hot metal warms up and rises | |
> | The hot metal sends infrared rays into the room | |

Staying cool by getting rid of heat

Water cooling

Car engines overheat if they are not cooled. Water is usually used and the water itself is cooled by passing it through a radiator. The water is pumped through narrow pipes that are joined to black cooling fins. These are then cooled by the air flow from a fan and by radiating a little heat. The cool water then returns to the engine block.

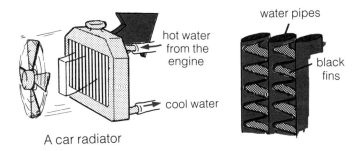

A car radiator

Air cooling

These are found on air-cooled engines, at the back of refrigerators and in car radiators. Their job is to get rid of heat. The fins make a large area of hot metal for the cooling air to flow over. The fins also radiate heat.

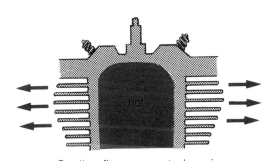

Cooling fins on a petrol engine

> 12 Why are cooling fins usually painted black?
>
> 13 Explain how conduction, convection and radiation take part when cooling fins cool a hot engine.

14 Copy this simplified diagram of how water is used to cool the engine of a car. Label these parts: radiator, fan, water pump, cylinder block. Use colours to show cool and hot water and arrows to show the direction of circulation.

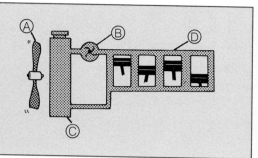

Staying cool by keeping heat out

Tanks that store chemicals that must not get too hot, are often painted silver and have double walls. Some are spherical to give them as small an area as possible. Fire-proof suits and space suits are made of shiny insulating material to stop heat from getting inside.

15 Put the following into the table below:
Cavity walls, fins on an engine, silver space suit, car radiator, house radiator, silver teapot, silver cold liquid storage tank, slippers, vacuum flask.

Designed to keep heat out	Designed to keep heat in	Designed to get rid of heat

Is a shiny surface used in these examples because it is a good reflector or a bad radiator?

19 Temperature

We often say 'it is hot today' or 'this tea is cold'. The scientific word for the level of hotness of things is **temperature**. It is important to know the temperature (hotness) of things like water, the air outside, our bodies or an oven. We have a sense that tells us when things are too hot or too cold, but can we trust our senses?

It is important to know temperature

Measuring temperature

EXPERIMENT

Put one finger into hot water and another finger into cold water. After 1 minute put both fingers into some warm water. Both fingers are now at the same temperature but do they feel the same temperature? The skin can be 'conditioned' and cannot be trusted to measure temperature properly. A swimming pool feels cold when you first dip your toe in, but feels warm after you have been in the water for a while.

Can we trust our sense of hot and cold?

1 **Thermometers** are used to measure temperature. This is because:
  (a) they are expensive;
  (b) they can be relied on not to change the reading as they 'get used' to the temperature;
  (c) they give a number reading;
  (d) they can measure temperatures too hot to touch;
  (e) they break easily.
  Write down the correct reasons.

Thermometers are marked to read temperatures in degrees Celsius (°C). Pure melting ice is always at the same temperature. This temperature is called 0 °C (the lower fixed point). The steam above boiling water under normal pressure is always at a fixed temperature too. This temperature is called 100 °C (the upper fixed point). These two temperatures are marked on a thermometer and the distance between the marks is divided into 100 degrees.

EXPERIMENT
Take a thermometer and check its fixed points (0 °C and 100 °C). Then use it to measure other interesting temperatures.

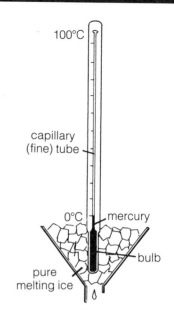

Checking the lower and upper fixed points of a thermometer

2 Alcohol (ethanol), usually coloured red, is sometimes used as the liquid in thermometers. Alcohol boils at 78 °C and freezes at −115 °C, and mercury boils at 357 °C and freezes at −39 °C. Alcohol expands much more than mercury and is cheaper. Which liquid would be the best to use in these thermometers?

Copy the table and insert 'mercury' or 'alcohol' into the correct box.

(a) A laboratory thermometer that is often used to measure the temperature of steam.	
(b) A cheap household thermometer that is used to measure the temperature of the air in a house.	
(c) A thermometer used to measure very low temperatures	
(d) A very sensitive thermometer that measures small changes in temperature	

3 Which of these properties of mercury make it a good liquid to use in thermometers? Copy the list and put 'good' or 'bad' against each property.

(a) Mercury is a good conductor of heat.	
(b) Mercury expands evenly as the temperature rises.	
(c) Mercury boils at 357 °C and freezes at −39°C.	
(d) Mercury is poisonous.	
(e) Mercury is expensive.	

Clinical thermometers

These are used to measure body temperatures.

EXPERIMENT. Using a clinical thermometer

1. Clean the thermometer bulb in an antiseptic solution.

2. Flick the mercury down into the bulb. Look carefully at the capillary tube and you will see it becomes very narrow indeed just above the bulb. The mercury can force its way up through this narrow section but cannot get back when it cools. The mercury is returned to the bulb by a flick of the wrist.

3. Place the thermometer under your tongue for one minute.

4. Take the thermometer out of your mouth and read the temperature. You will notice that, because of the constriction, the mercury stays up the tube and the thermometer can be read away from the patient. The mercury thread is very fine to make the thermometer sensitive. Often the glass stem of the thermometer is triangular and magnifies the mercury thread when held at the correct angle – it does not roll so easily either.

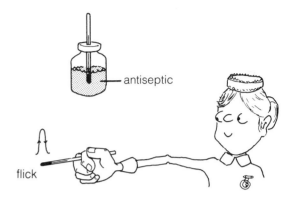

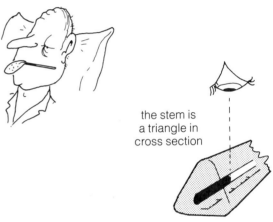

the stem is a triangle in cross section

4. Why does a clinical thermometer only measure temperatures from about 35 °C to 42 °C?

5. What are the advantages of a very fine capillary tube?

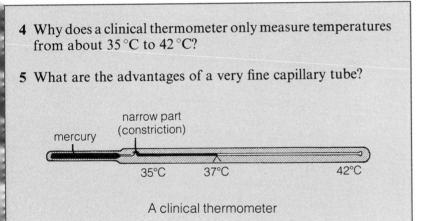

A clinical thermometer

Electronic thermometers

These use a probe to convert temperature into a voltage and electronics to display this voltage as a digital read-out:

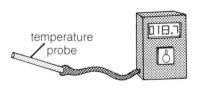

An electronic thermometer

6. Separate these remarks about electronic thermometers into advantages and disadvantages: easy to read, very expensive, bulky, robust, can measure from about −50 °C to 200 °C, need batteries, do not contain poisonous liquid, small probe can reach awkward places.

Heat energy check list

After studying the section on Heat energy you should:

- know that heat is a form of energy that flows from hot to cold bodies
- know how to convert heat energy into other forms
- understand what a heat engine does
- understand broadly how the following heat engines work: steam turbine, jet engine, rocket engine, petrol engine
- know that heat makes materials expand
- know of problems caused by expansion and how they are overcome
- know how a bimetal strip and thermostat work
- know that liquids expand more than solids
- understand how heat is transferred by convection
- understand your hot and cold water system
- know that gases expand more than solids
- know of things caused by the convection of air
- understand what is meant by conduction of heat
- know that water and air are bad conductors of heat
- know of uses of good and bad conductors
- understand that heat can travel as electromagnetic waves
- know some properties of heat radiation
- know that a black surface is good at absorbing and radiating heat waves . . .
- . . . and that a shiny surface is the opposite
- understand why a vacuum flask keeps things hot
- understand the main ways that a house loses heat . . .
- . . . and how to reduce those loses
- know methods of keeping cool
- understand the idea of temperature
- know what is meant by the fixed points
- know how to use mercury, clinical and electronic thermometers

Revision quiz

Use these questions to help you revise the section on Heat energy.

What is heat?	... it is energy that flows from a hot body to a cooler one.
How can you convert heat energy into (a) electricity? (b) into movement of a wheel?	... with a thermocouple. ... by a turbine (which can turn a generator to make electricity).
Name an engine that changes the energy of hot steam into electricity.	... steam turbine and generator.
Name a heat engine that uses hot gas to drive an airliner.	... a gas turbine (or jet engine).
Name a heat engine used to drive cars.	... a petrol engine (or diesel engine).
Name a heat engine that can be used in space	... a rocket engine.
How are the following problems overcome: (a) the expansion of concrete roads; (b) the expansion of metal or concrete bridges; (c) pouring boiling water into glass?	... by leaving gaps between concrete slabs. ... by leaving the ends free to move. ... by using Pyrex glass that expands very little.
What is a bimetal strip?	... a strip made of two different metals welded together.
Why does it bend when heated?	... because one metal expands more than the other.
Which metal is on the outside of the bend when heated?	... the metal that expands the most.
What is a thermostat?	... an automatic switch that is controlled by temperature.
How does the thermostat in an electric iron work?	... it goes on when the iron is cold and off when the iron is hot enough.
How does the thermostat in a fridge work?	... it goes on when the temperature is too high and off when it is low enough.
How does a thermostat work?	... electric current passes through a bimetal strip and two contacts. When the temperature changes the strip bends and makes or breaks the contacts.
What is meant by the convection of air?	... it is the upward movement of hot light air caused by the downward movement of heavy cold air.
Name five things that are caused by the convection of air.	... smoke rises to the top of a room, cold air falls to the bottom of a fridge, hot air balloons fly up, skirts keep your legs warm, birds and hang-gliders can soar.
Why is cotton wool such a good insulator?	... because pockets of air are trapped between the fibres and air is a very bad conductor of heat.
Name three other good insulators that use trapped air pockets.	... feathers on birds (or in duvets), woolly jumpers, plastic foam.
Name four properties of heat radiation	... it is electromagnetic, travels at the speed of light, can be reflected and focused, can travel through a vacuum.
What type of surface is good at absorbing and radiating heat radiation?	... a rough dark or black surface is a good absorber and radiator of heat radiation.

- What type of surface is bad at absorbing and radiating heat radiation?

　. . . a shiny polished surface.

- How does a vacuum flask reduce heat loss?

　. . . the cork reduces convection, the vacuum reduces conduction and the silvering reduces loss by radiation.

- Name three ways of reducing heat loss from a house.

　. . . insulate the loft, double-glaze the windows, use cavity walls, stop draughts through doors and windows.

- How do cooling fins on engines work?

　. . . they increase the area of hot metal in contact with the air to increase convection loss. They are sometimes black to increase radiation loss.

- What is the ice point?

　. . . it is the temperature of pure melting ice (0 degrees on the Celsius scale).

- What is the steam point?

　. . . the temperature of pure boiling water at standard pressure (100 degrees on the Celsius scale).

- Why are thermometers better than skin for measuring temperatures?

　. . . they give a number reading and skin gives us different feelings as it 'gets used' to a temperature.

Examination questions

(a) Fig. 1 shows a modern hot water cylinder. It has a volume of 200 litres and contains two immersion heaters labelled P and Q.
  (i) Write down the approximate volume of water which element P can heat up.
   Show how you got your answer.
  (ii) Electricity is available either at a cheap 'off peak' rate or a more expensive rate.
   Explain why
   (A) Q is used during the off peak period,
   (B) P is necessary.
(b) The temperature of the water is controlled by a device called a thermostat. Fig. 2 shows a thermostat.
   X and Y form a bimetallic strip
   Z is an electrical insulator.
  (i) Name a suitable material for Z.
  (ii) Explain why the bimetallic strip bends when it becomes hot.
  (iii) Explain what effect this has on the current through the heater.
  (iv) If the temperature control screw is turned so that it moves to the left, does the temperature of the water increase, decrease or stay the same?

(NEA)

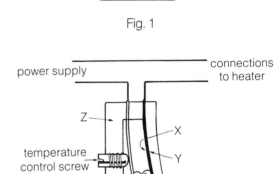

Fig. 1

Fig. 2

(a) (i) Name the three ways by which heat energy can be transferred.
  (ii) Which is the *main* method of heat transfer in each of the following cases?
   (A) heat energy reaching us from the sun;
   (B) heating water in an electric kettle;
   (C) heating a room using central heating radiators.
(b) Fig. 3 shows a pudding which has been cooked in an oven at 200 °C for a time of fifteen minutes.

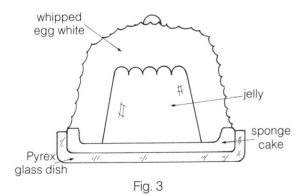

Fig. 3

  (i) Explain why the jelly has not melted.
  (ii) The oven elements have a total power of 2 kW. The elements are switched on for a total time of 15 minutes. If the cost of electricity is 6p per kilowatt-hour, calculate the cost of cooking the pudding.

(NEA)

3 This question is about heating and temperature change.
An electrical immersion heater is used to heat up some water in a beaker. The water is stirred all the time and its temperature is measured every minute.
Fig. 4 shows how the energy output of the heater increases with time.
Fig. 5 shows how the temperature of the water changes with time.
  (a) Using Fig. 4 find the energy output of the heater in
      (i)   2 minutes,   energy output = ...
      (ii) 5 minutes,   energy output = ...
      (iii) 10 minutes,   energy output = ...
  (b) Using Fig. 5 find the TEMPERATURE RISE in K (°C) of the water in
      (i)   2 minutes,   temperature rise = ...
      (ii) 5 minutes,   temperature rise = ...
      (iii) 10 minutes,   temperature rise = ...
  (c) Complete the table using your answers to (a) and (b).

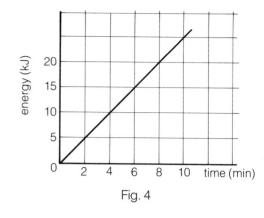

Fig. 4

Time in minutes	0	1	2	5	10
Energy in kJ	0				
Temperature rise in K	0				

  (d) Why is the temperature rise in the first 5 minutes greater than the temperature rise in the second 5 minutes? Write TWO reasons.
  (e) Use Fig. 4 to calculate the power output of the heater. Show your working.

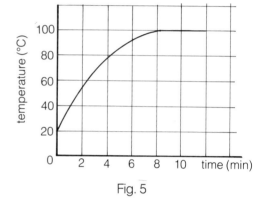

Fig. 5

4 (a) Why do air-filled cavity walls keep a house warmer in winter than solid brick walls?
  (b) Why does filling the cavity with plastic foam keep the house even warmer?
  (c) Explain how a hot water radiator heats a room.

(MEG)

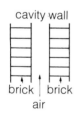 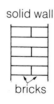

Fig. 6

5 The diagram shows three glass containers, A, B and C, in a water bath.
The water bath is gently heated.
  (a) Which tube will overflow first?
      Explain your answer.
  (b) Which tube will be the last to overflow?
      Explain your answer.

(MEG)

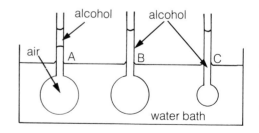

Fig. 7

176

Fig. 8 illustrates an instrument used to measure the time that the sun shines during a day. The blackened glass bulb contains mercury and is supported inside an evacuated glass case. Fig. 9 shows how the connecting wires are arranged inside tube A.
(a) How does energy from the sun reach the mercury? Give a reason for your answer.
(b) Explain why the clock starts when the sun shines.
(c) Why is tube A of small cross-sectional area?
(d) Explain why blackening the bulb ensures that the mercury level falls rapidly when the sun ceases to shine.

(MEG)

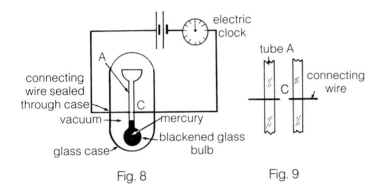

Fig. 8   Fig. 9

(a) The amount of energy required to heat a house depends on its size and how well it is insulated. The main areas of heat loss are losses through (i) the cavity walls, (ii) the windows and doors, (iii) the floor, and (iv) the roof.
How could each of these losses be reduced? In your answers you should describe the type of materials used to reduce losses and explain why they have improved insulation.
(b) It has been estimated that the Earth receives from the sun 10^{17} joules of energy per second and that on average 10^{14} joules of energy per second are being used on Earth, most of which comes from burning fossil fuels. If the earth is receiving 1000 times as much energy as we use, and it is free, why do we not make use of this energy rather than burn so much fossil fuel?
(c) The diagram above shows, in simplified form, one method of heating a house using solar energy. It consists of a roof-mounted heat-absorbing panel filled with water.
  (i) Explain how the system works, clearly stating the main energy change that takes place.
  (ii) Why is the absorbing panel usually painted black?
  (iii) Suggest ONE reason why the transparent cover increases the efficiency of the system.
  (iv) Explain the purpose of the expanded polystyrene board.
  (v) Give TWO reasons why this system cannot be used as the sole supply of domestic heat in Britain.

(NISEC)

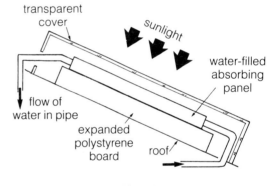

Fig. 10

20 Electric energy

Electric charge

Try these experiments if you can.

Rub a piece of dry polythene energetically on a jumper or cloth. Place the end you rubbed close to:
(a) a feather,
(b) some small pieces of paper (or other light objects),
(c) a thin stream of water.

Rub a balloon against a woolly jumper and drop pieces of fluff, paper, soap bubbles or feathers past the places you have rubbed.

Make a balloon stick to your clothes (or the ceiling) by rubbing it on a cloth.

Cut out two plastic penguins from biscuit wrappers, rub them with a cloth and bring them close together.

These experiments show that when you rub pieces of plastic with a cloth, they get the power to move light objects. The plastic picks up an electric charge and becomes charged with electricity.

The electric charge on each piece of plastic gives it energy. This energy is stored on the plastic for a while and so is called **electric potential energy**.

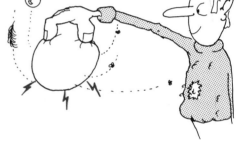

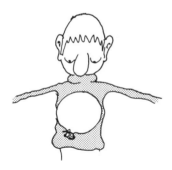 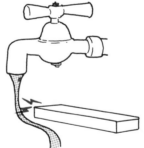 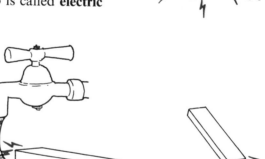

---

1 What form of energy was used to make the electric potential energy on the balloon. What other form of energy was made at the same time? Copy and label the energy arrow.

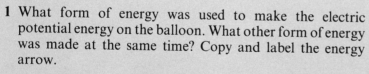 

(Electrical, Heat, Kinetic)

EXPERIMENT. Getting a shock from electric charge

Press a lump of plasticine onto a large metal tray. Using the plasticine as a handle, rub the tray over a sheet of plastic (a plastic rubbish sack will do). Lift the tray and bring it near to your body. You should see and feel sparks. If you have a small neon lamp, hold one of its wires and touch the other on the tray. The lamp will flash as electric charge passes through it to your hand.

Getting a shock from electric charge

Lighting a lamp with electric charge

2. The tray picks up electric charge from the plastic sheet and gets electric potential energy. A spark has heat, light and a little sound energy. Draw an energy arrow that shows the changes that take place when you get an electric shock from the tray.

More about electric charge

EXPERIMENT. To show that when two materials are rubbed together, they both become charged

It is best to do this experiment in the dark.

Use the tray and plastic bag of the last experiment. Charge up the tray and test it with a neon lamp (the lamp will flash). Then test the part of the plastic bag that was underneath the tray. Touch the wire of the neon lamp on the plastic. You will see it glimmer with light as you move it from place to place. When you charge up the tray you also charge up the plastic bag.

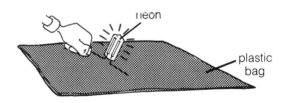

The bag and the tray are both charged

3. In this experiment the tray and the plastic bag become charged and gain electric potential energy. (They both make the lamp light.) Where does that energy come from?

4. Touching the tray with the neon lamp gives one big flash of light. Touching the plastic bag in different places gives lots of little flashes. Can you explain why the tray (metal) and plastic bag behave differently?

There are two types of electric charge

An electrometer is an instrument that measures electric charge. Rub a polythene rod and touch it on the cap of an electrometer. You will notice that the pointer of the meter deflects. If you then rub a piece of cellulose acetate (the type of plastic used to make film) and place that on the cap, the meter deflects the other way. Clearly the charges on the rods are not the same. There are two types of electric charge. Both types can pick up light objects and cause sparks. They are called positive and negative. (No one has yet discovered a third type of electric charge.)

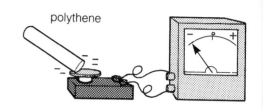

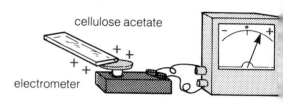

Using an electrometer to show there are two types of charge

> **5** When the charged rods are removed from the electrometer, the pointer does not go back to zero. Can you explain why?

Forces between charges

EXPERIMENT

You can see that there are two types of charge with this experiment. It also shows the forces that act between them.

Charge up polythene and acetate rods and balance them on watch glasses so that they can spin freely. Then bring up a charged polythene rod to each in turn. In one case the rods attract and in the other they repel. So clearly the charges on the polythene and acetate are different.

Next bring up a charged acetate rod to each rod and watch for movement between the charges.

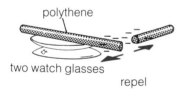
repel

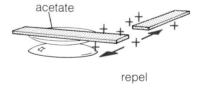
repel

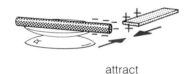
attract

Testing the forces between charges

> **6** Copy and put 'attract' or 'repel' in the boxes to describe the type of force between the charged rods.
>
> | | Polythene | Acetate |
> |---|---|---|
> | Polythene | | |
> | Acetate | | |

7 Copy and say whether these charges attract or repel when they meet.

8 Copy and complete this summary of the way forces act between electric charges:
like charges . . . .
unlike charges . . . . .

	Negative charge	Positive charge
Negative charge		
Positive charge		

Insulators

The charges on the plastic rods stay in place. They do not move along. Plastic and other materials that do not allow electric charge to move are called **insulators**. Electricity that does not move is called **static electricity**.

Conductors

These are materials, such as metals, through which electric charge can move freely. These are called **conductors** of electricity.

9 If a charged <u>metal tray</u> is held by a <u>plasticine</u> handle, the charge remains on the tray. But if the tray is touched by a <u>person</u> on the <u>ground</u>, all of the charge runs away. Say whether each of the items underlined is an insulator or a conductor of electricity. Would a charged plastic tray lose all of its charge if you touched it like this?

Storing electric charge

Capacitors

Capacitors are electronic components that are used to store charge or separate a.c. from d.c. One common type is made from two thin aluminium plates separated by an insulator such as waxy paper. This sandwich is rolled into a small cylinder.

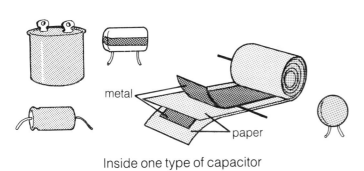

Inside one type of capacitor

EXPERIMENT. Capacitors can store electric charge

Connect a battery to a capacitor. Current flows for a moment as the plates become charged – one positive, the other negative. Take away the battery and connect the charged capacitor to a lamp. The lamp lights for a moment as charge from the capacitor flows through it.

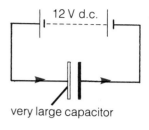

(a) Charging a capacitor

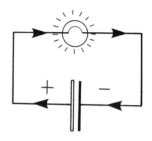

(b) Discharging a capacitor

10 What energy changes take place in circuit (b) when the capacitor discharges?

11 Does the 'discharge current' (in circuit (b)) flow the same way as the charging current or the opposite way?

12 What can be 'stored' in the following items: sponge; a purse; storage heaters; a toilet cistern; a record; a capacitor?

Electric potential energy is measured in volts

A gold-leaf electroscope is useful for this investigation. It uses a strip of gold leaf to detect electric charge. The leaf is very light and is fixed at one end to a metal bar. When charge is placed on or near the cap of the instrument, the leaf rises.

Put some positive charge on the electroscope by scraping a charged acetate rod across its cap. The leaf will open showing that the leaf has become charged and has electric potential energy.

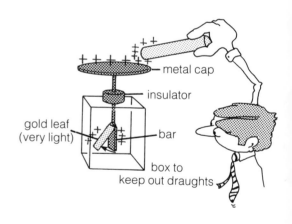

Next try to give the leaf the same electric potential energy by connecting it to a battery. You will have to use a voltage supply that can produce several thousand volts to make the leaf open by the same amount.

The small charge on the leaf gives it an electric potential of several thousand volts. Electric potential is measured in volts.

Positive and negative electric energy

The positive charge on the leaf gave it an electric potential of (say) +2000 volts. A negative charge on the leaf could give it a negative energy of (say) −2000 volts. Both kinds of energy can cause sparks and other events to happen.

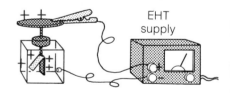

Things that have no charge, such as the Earth, have an electric potential of 0 volts.

13 Copy and put 'volts', 'electric charge' and 'electric potential energy' in the empty boxes.

What is given to an object when it is rubbed?	
What is the form of energy that a charged object has?	
What is the unit used to measure electric potential?	

Earthing

A charged electroscope may have an electric potential of about 2000 volts. At voltages like this, the earth and your body are conductors of electricity. When you touch the cap, the positive charge on the leaf and cap moves through your body to the ground. The net charge (and the electric potential) of the electroscope becomes zero when the apparatus is earthed.

Using very large charges

A Van de Graaff generator is a machine that can produce large charges and very high voltages. A rubber belt carries negative charge up to a large metal dome. The dome is supported on an insulating column so that the charge on it cannot escape. If you charge the dome and place an 'earthed' ball nearby, large sparks, like miniature lightning, strike across the gap. There is so much negative charge on the dome that the air (normally an insulator) becomes a conductor for a moment. There is a spark and the charge escapes to the 'earthed' ball (or you if you get too close).

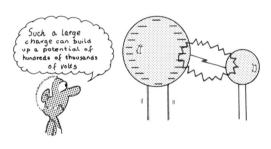

14 A spark changes electric potential energy into other forms. Copy and label the energy change arrow for a spark.

Sparks can be dangerous

Place a sheet of paper in the path of sparks from an induction coil and you may see the paper burst into flames. Electric sparks are hot.

Sparks from static electricity are hot enough to make a mixture of gas and air burst into a flame. If you make sparks jump from a charged metal tray to the nozzle of a gas burner, the gas from it may light. A tiny spark can produce a frightening burst or flame.

Dangerous situations

The flame from a gas burner burns steadily but some mixtures of gas and air do not burn. They explode violently. Examples of these explosive mixtures are petrol vapour and air and some anaesthetics. A small spark can trigger off a large explosion of these mixtures.

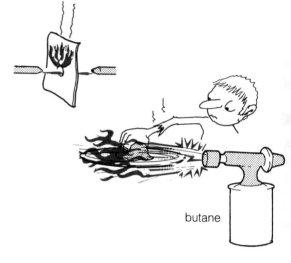

1. Petrol and oil tankers

The movement of petrol and oil as it is being transported in tankers can build up an electric charge. A spark from this charge could make petrol vapour explode. An escape route for the charge has to be provided to prevent any possibility of sparking. A chain to the ground, or tyres made from special conducting rubber, are used to allow the charge to escape to 'earth'.

2. The operating theatre.

Sudden movement of blankets, apparatus or clothes in operating theatres can cause sparks to fly. Careful precautions are taken to make conducting paths to the ground so that static charges do not build up. Sparks are especially dangerous because of the flammable gases used to anaesthetize the patient and the large amount of oxygen present in the air of the theatre.

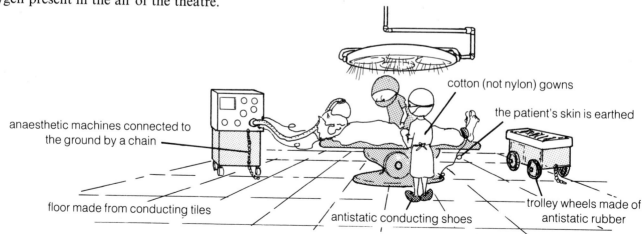

3. Lightning.
Thunderclouds are natural electrostatic generators. Violent activity inside the cloud separates electric charge making the top of the cloud (+) and the bottom (−). When the voltage is high enough, lightning strikes, either from the bottom of the cloud to the ground or between clouds. The energy of the lightning (several thousand 'units') and the temperature of this giant spark (about 10 000 °C) can cause considerable damage. You would be unlikely to survive a direct hit by lightning.

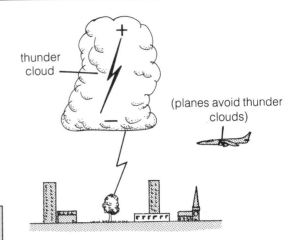

15 Write an essay on the dangers of static electricity.

Charges, atoms and electrons

We are familiar with the idea that all matter is made of atoms and these atoms are extremely small (see p. 314). For many years it was thought that atoms were the smallest particles in existence. However, in 1897 an even smaller particle was discovered. This particle had a negative charge and much less mass than an atom. Atoms are small but this particle was two thousand times smaller than the lightest atom. It is now thought to be one of the particles found inside an atom. Its name is the **electron**. Work with radioactive substances (p. 320) led to the discovery that atoms have a very small, dense **nucleus** at the centre with a positive charge. An atom is made from a nucleus and electrons. Although the nucleus is much heavier than the electrons it has the same amount of electric charge. The atom as a whole is not charged because the (+) charge on the nucleus and the (−) charge on the electrons cancel out. (An atom can sometimes lose an electron, in which case the atom will have a net (+) charge.)

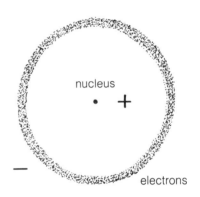

A picture of an atom

The electron theory of electric charge

When polythene is charged by rubbing it with fur, electrons move from the fur to the polythene. Since electrons are charged negative, the polythene gets a negative charge and the cloth is left with a positive charge.

185

16 When fur charges polythene one of the materials gets slightly heavier. Which one is this?

17 Copy and complete.

Materials that are rubbed together	Type of charge it gets	Has it lost or gained electrons?
Polythene		
Fur		
Cellulose acetate	positive	
Cloth		

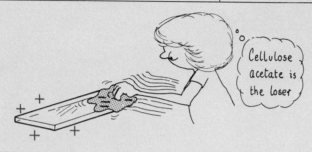

Inducing an electric charge

EXPERIMENT. A charged rod attracts uncharged objects

Move a charged rod towards a small piece of uncharged aluminium foil. As the charge approaches, the foil will begin to move and may even be lifted off the ground.

Why the foil moves

The nearby negative charge on the rod repels the negative electrons in the aluminium. They move through the metal to its furthest surface and leave the top surface with a positive charge. In this way equal and opposite charges are forced (or induced) onto the surfaces of the aluminium by the charged rod.

The induced charges are then attracted and repelled by the charge on the rod. Since the attraction of the + charge is greater than the repulsion of the − charge (which is further away), there is a net force of attraction that can lift the foil.

Induced charges also appear on insulators such as paper and dust. Their electrons are not free to move but their molecules become distorted and charges appear on the top and bottom surfaces. So they too experience a net attractive force towards charged bodies.

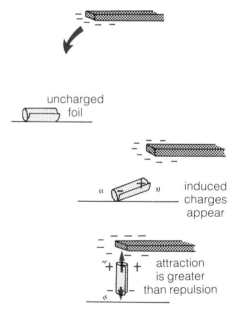

18 Explain why a small piece of uncharged aluminium foil is attracted to a charged polythene rod and why sometimes it flies away after it has made contact.

Electric charge on the move

There are two sorts of electricity: static electricity produced by friction and electric current produced by batteries and generators. What is the link between these two forms of electricity?

EXPERIMENT. Travelling electrons

Connect a thread from the dome of a Van de Graaff generator to a very sensitive ammeter. Complete the circuit with a wire from the meter to the base of the generator. Turn the belt and charge up the dome of the generator. After a short delay, the meter should deflect.

A negative charge on the dome means it has a large number of extra electrons. The thread is an escape route for these electrons so they travel down it and through the meter. As the electrons pass through the meter they make the pointer deflect.

Next connect a battery and large resistor to the same meter. A small electric current flows through the meter and deflects it in the same way as before.

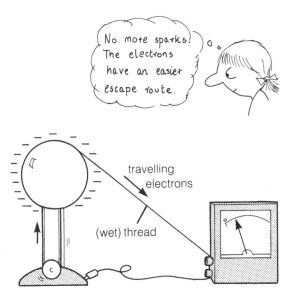

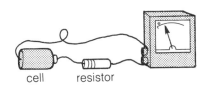

The moving electrons and the electric current have the same effect on the meter. This suggests that an electric current is the movement of electrons.

Lamps without filaments also show that electron flow is the same as electric current

Fluorescent tube lights and the brightly coloured 'neon' signs used by advertisers are lamps that have no filament. The electricity passes through a gas inside a tube. Small neon lamps are examples of these gas discharge lamps. They glow with red light when the voltage is high enough (about 50 volts) to pass current through the neon gas inside the lamp. If electrons from a Van de Graaff dome pass through such lamps they also glow with light. Electric current and moving electrons have the same effect when they pass through the neon gas.

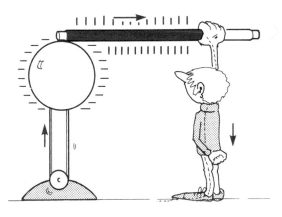

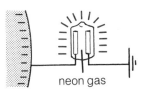
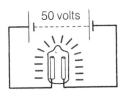

Electric current in wires

A copper wire contains many millions of electrons. If the ends of the wire are made positive and negative (by connecting to a battery for example) some of the electrons begin to drift to the end that is positive. This movement of electrons is an electric current.

Marathon fever – a current of people on the move

When the gates are opened, the great crowd of runners begins to steam along the race route – a jostling current of people.

The runners are rather like the great crowd of electrons on the dome of a Van de Graaff machine. When a thread is connected to earth, the electrons begin to steam along it – a current of electrons or electric current.

Coulombs and amperes

These are the units used to measure electric charge and electric current.

Electric charge is measured in **coulombs**.
This is a large charge and it would take about 10^{18} electrons to make 1 coulomb of negative charge.

> **19** About how many electrons would make 1 microcoulomb of charge?

Electric current is measured in **amperes** (or amps for short).

Electric current is the flow of electric charge and so the coulomb and the ampere are connected.
In fact:

> one coulomb = the amount of charge that passes when one amp flows for one second

The word equation connecting current, time and the amount of charge that passes is:

> charge   = current × time
> (coulombs) = (amps)   × (seconds)

> **20** Copy and put in the units used to measure each of the quantities.
>
> | | |
> |---|---|
> | Electric charge | |
> | Electric potential energy | |
> | Electric current | |

21 Use the equation to calculate the unknowns in this table.

The size of the current (amps)	The time it flows	The charge that passes (coulombs)
2	10 s	
0.5	1 h	
2.5		1000
	32 s	256

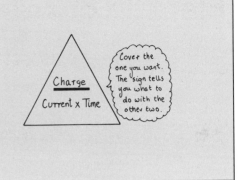

Electrons in space

Metal wires contain millions of electrons. Can these electrons be got out of the wire?

Simply heating a wire gives some electrons enough energy to 'escape' from the metal. Electrons are released when the filament of a lamp is heated for example. But the electrons get mixed up with the gas inside the lamp and nearly all of them return to the filament.

Special glass tubes, from which the air has been removed, have been designed to use these freed electrons.

Electron beams

Electron beams are electric currents in space – without wires.

Electron beams in vacuum tubes are essential parts of television sets, oscilloscopes and radar equipment. This diagram shows a vacuum tube that produces a fine beam of electrons. A metal cylinder with a hole in the centre is placed in the neck of the tube.

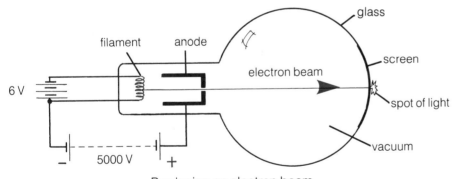

Producing an electron beam

The cylinder, called the anode, is given a high positive voltage. This attracts electrons released by the filament and accelerates them to enormous speeds. Some electrons pass through the hole and into the vacuum space – a fine, high-speed beam of negative particles. The tube has a screen painted on the inside of the glass and conducting paint that leads back to the anode lead. Where electrons hit this screen, a spot of light is seen.

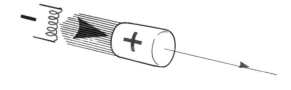

22 How are the electrons got out of the filament and into the vacuum tube?

23 How are the electrons got moving and formed into a fine beam?

24 Why must there be a high vacuum inside the tube?

25 When the electrons smash into the glass, why do they not break it?

26 The paint that is used to make the screen contains a chemical that 'fluoresces'. What does this mean?

27 What energy changes take place when the electrons hit the fluorescent screen?

28 If the filament is 'boiling off' electrons all the time, why does it not boil dry? Why does it not run out of electrons?

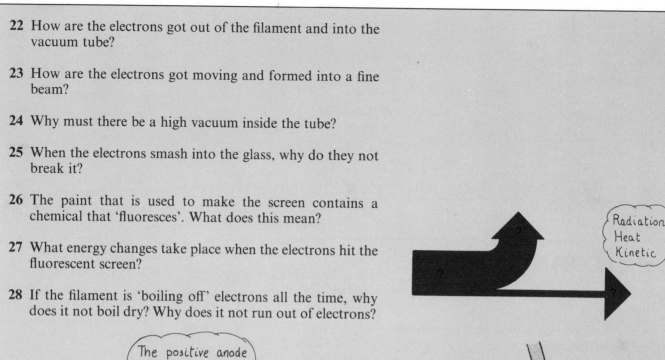

The control of electron beams

We must be able to deflect electron beams to make use of them.

Electrostatic deflection

A vacuum tube is needed with two metal plates inside as shown. A high-voltage battery is connected to these plates making one positive and the other negative. As electrons pass between the plates they are attracted by the positive plate and repelled by the negative. They move down as they pass and the spot shifts downwards (the electrons are moving too fast to hit the positive plate).

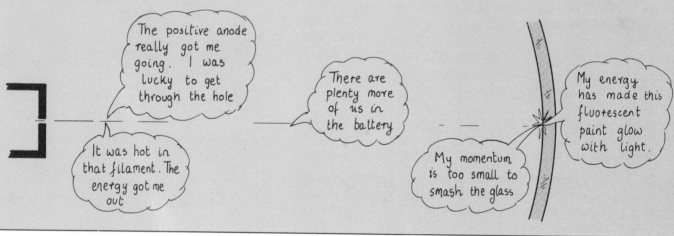

Electrostatic deflection of an electron beam

29 How would you use the same plates to deflect the beam upwards?

30 What would you see happening to the beam if an alternating voltage was put on the plates?

Horizontal plates, like those that deflect the beam up and down are called Y-plates. Some tubes have two vertical plates called X-plates fitted as well. Voltages on these plates will deflect the beam sideways. Oscilloscope tubes have X- and Y-plates.

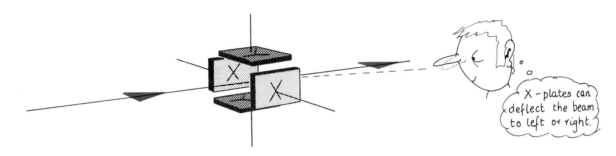

The oscilloscope

The oscilloscope is a useful instrument for giving a 'picture' of voltage, especially voltage that changes quickly. It has a vacuum tube with a filament and anode to produce an electron beam, and two sets of deflecting plates (X and Y).

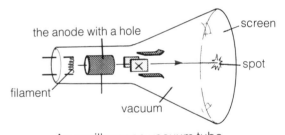

An oscilloscope vacuum tube

EXPERIMENT. Using the Y-plates

Switch on an oscilloscope and adjust it until you have a spot in the centre of the screen. The Y-plates are usually connected to the two terminals of the oscilloscope.

(a) Connect a dry cell to the two terminals. What happens to the spot? What happens to the spot if the cell is connected the other way round? Can you explain why the spot jumps up and down?

(b) Connect an alternating voltage to the Y-plates (e.g. from a transformer or an alternator). You will see a vertical line appear on the screen. The spot travels up and down tracing out a line of light. The spot moves so fast that the light has not time to die away before a fresh line is drawn.

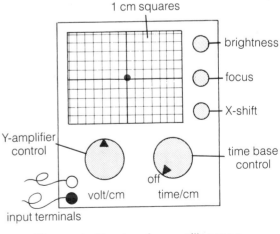

The control knobs of an oscilloscope

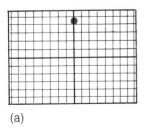
(a)

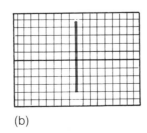
(b)

191

The Y-amplifier

The oscilloscope has an amplifier connected to the Y-plates that can magnify the size of the voltage that is being examined. This is especially useful if the voltage is small (like the voltage from a microphone). The Y-amplifier control knob shows how many volts are needed for a deflection of 1 cm.

EXPERIMENT. Using the X-plates

The X-plates of an oscilloscope are usually connected to an electronic circuit that pulls the spot steadily across the screen from left to right. When the spot reaches the end of the screen it flicks back quickly and starts again. This circuit is called the **time-base circuit**. The time-base control knob can change the time the spot takes to cross the screen. Turn on the time base and watch the spot cross the screen (the fly-back is too fast to be seen). Notice that when the spot moves quickly it draws a steady line of light.

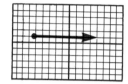

EXPERIMENT. Using the X-plates and Y-plates at the same time

Connect a slow alternating voltage to the Y-plates of an oscilloscope (an alternator being turned once a second will do) and watch the spot as it moves up and down on the screen. Switch on the time base so that as the spot goes up and down, it goes sideways as well. These two movements together draw a wave on the screen.

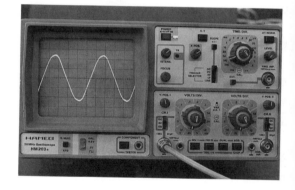

31 If the time taken to go up and down and the time taken to go across are both one second, then one wave is drawn. How many waves would be drawn if the time to cross was changed to 2 s?

Replace the slow alternating voltage with one of mains frequency (50 Hz) from a transformer. Adjust the time base until one wave appears 'frozen' on the screen. The spot draws out 50 waves exactly on top of each other in one second and a wavy line of light is produced.

32 This diagram shows a section through an oscilloscope.
   (a) What happens at the filament when switch K is closed? State the energy changes that take place then.
   (b) How does the high voltage on the anode help to produce the electron beam?
   (c) Explain why the tube must be evacuated.
   (d) Explain what will happen to the electron beam if plate K is given a positive charge and plate L a negative charge.
   (e) Draw a sketch of the oscilloscope screen showing:
     (i) an alternating voltage waveform.
     (ii) an alternating voltage of twice the frequency of (i).

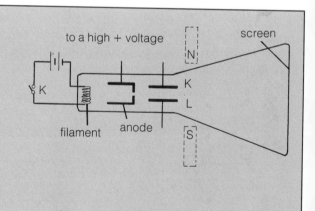

Electric charge check list

After studying the section on Electric charge you should know:

- there are two types of electric charge, positive and negative
- that positive charge cannot be produced without an equal amount of negative charge
- that negative charge is due to the presence of extra electrons
- that charges exert a force through space on other charges
- that like charges repel and unlike charges attract
- that electric charge can move through conductors but not through insulators
- that + and − electric charge can be stored in a capacitor
- that a charged body has electric potential energy, measured in volts
- situations where sparks can be dangerous
- that a charge can be induced on objects by a nearby charge
- that earthing means connecting a body to the ground by a conductor
- that the earth and all earthed bodies have an electric potential of 0 volts
- that an electric current is the movement of electric charge
- that current is measured in amperes and charge in coulombs
- that one coulomb is the amount of charge that passes when a current of 1 amp flows for one second.
- that electrons can be released from a wire by heating it.
- that electrons can travel through a vacuum and be accelerated by a + charge
- that an oscilloscope uses an electron beam to draw patterns on a screen and uses charged plates to deflect them

Revision quiz

Use these questions to help you revise the work on Electric charge.

- How can you charge a piece of plastic?
  ... by rubbing it with a dry cloth.

- If electrons are transferred from the cloth to the plastic, what charge will each have?
  ... the cloth will have a positive charge and the plastic negative.

- What is the law of force between charges?
  ... like charges repel, unlike charges attract.

- What sort of energy does the charged rod have?
  ... electric potential energy.

- What unit is used to measure this energy?
  ... joule/coulomb or volt.

- What is meant by negative electric potential energy?
  ... the energy of a body with a negative charge – it can cause movement and sparks like positive energy.

- What is a conductor?
  ... a material through which electric charge can move.

- What is an insulator?
  ... a material through which electric charge cannot move easily.

- What are induced charges?
  ... charges produced on the surface of a body by a nearby electric charge.

- Why does a charge attract an uncharged body?
  ... because the attraction to the opposite induced charge on the nearest part of the body is greater than the repulsion of the like charge on the furthest part of the body.

- What is a capacitor?
  ... it is an electronic component that can store electric charge.

- What is earthing?
  ... it is the connection of a body to the ground by a conducting wire. It gives the body an electric potential of 0 volts.

- What is an electric current?
  ... it is the flow of electric charge or electrons.

- What is the unit of electric current?
  ... current is measured in amperes (amps).

- What is a coulomb of charge?
  ... it is the amount of charge that passes when a current of 1 amp flows for 1 second.

- What is the equation connecting current and charge?
  ... charge = current × time.

- How can electrons be released from a metal?
  ... by heating it to a high temperature.

- How can electrons be made into a fine beam?
  ... by attracting them (in a vacuum) with a high voltage to an anode with a fine hole in its centre.

- How can an electron beam be deflected?
  ... by passing it between + and − charged metal plates.

- Name three devices that use electron beams
  ... oscilloscopes, TV sets, computer monitors.

21 Cells and voltage

The things in the photograph all need electric energy to work them. They use cells to provide that energy. Cells have two terminals called positive (+) and negative (−). The terminals of this common type of cell are the button on the top and the case at the bottom.

Chemical activity inside the cell keeps the + terminal full of positive charge and the − terminal full of negative charge. These charges give the terminals different electric potentials.

The difference between the energies of the terminals is called the cell's **potential difference (p.d.)**. Potential difference is measured in **volts** and is commonly called **voltage**.

The voltage of a cell is measured by a **voltmeter** in volts. Note how to connect the terminals of a voltmeter to a cell.

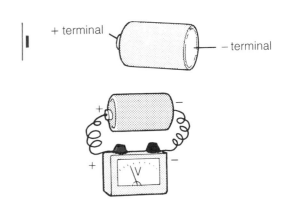

Using a voltmeter to measure the voltage of a cell

Electromotive force (e.m.f.)

As electric charge passes through a battery, it gains energy. This energy comes from chemical reactions inside the battery and is used by the charge as it passes round the outside circuit. The energy supplied to each coulomb is called the battery's **electromotive force (e.m.f.)**. e.m.f. is measured in volts and can be found by connecting a high resistance voltmeter to the battery when it is not supplying a current to other components.

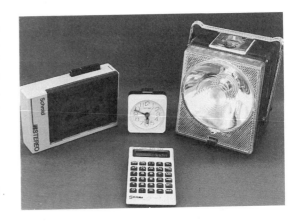

EXPERIMENT

Measure the voltage of two cells separately and joined together in the three ways shown. Record your results neatly.

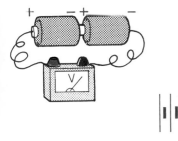 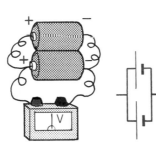 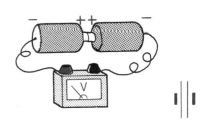

1 A student recorded her results in pictures like this. The numbers in the circles are the voltages when the cells are put together. Can you explain these voltages?

   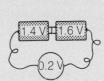 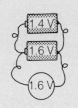

Batteries

A number of cells together is called a battery. The cells are connected + to − so that their voltages add up. The drawings show how the cells are arranged in some common batteries. Open up some old batteries if you can and find out how the cells are connected together.

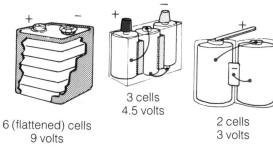

6 (flattened) cells
9 volts

3 cells
4.5 volts

2 cells
3 volts

The cells inside batteries

> **2** What is the link between the number of cells in each battery and its voltage?

Batteries and cells like these can only provide small electric currents and have to be thrown away when they are dead. For the larger currents needed by such things as cars and electric milk floats, batteries containing liquids are used. Lead–acid batteries are examples of these. This sort of battery is expensive but can be recharged and used again and again.

Lead-acid cells

EXPERIMENT. A home-made rechargeable cell
Put two clean lead plates into dilute sulphuric acid. Clip a voltmeter to the plates and check there is no voltage reading.

Connect a voltage supply to the two plates and allow it to drive current through the acid for 10 minutes. Disconnect the voltage supply and see if there is a voltage reading now. The cell should be charged and show a voltage.

Find out how long your lead–acid cell can light a small lamp by connecting the lamp to the plates. Time how long it takes the lamp to go out and for the cell to become flat.

You can then recharge the flat cell by connecting it to the voltage supply again.

Cells like this, that can be charged with energy from another supply, are called rechargeable cells.

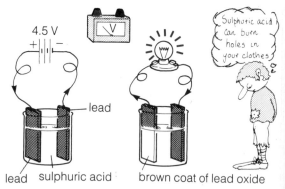

Charging and discharging a lead/acid cell

> **3** What can these rechargeable devices store up? Copy and complete.
>
> | Spring | Fountain pen | Water pistol | Lead–acid cell | Storage radiators | Sponge |
> |--------|--------------|--------------|----------------|-------------------|--------|
> | | | | | | |

4 When a lead–acid cell is charged, chemicals are formed on its plates that store the energy of the cell. Which of these diagrams shows the energy changes that take place while the cell is being charged, and which shows the discharge energy changes? Copy the diagrams and label them 'charge' and 'discharge'.

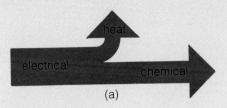

Car batteries

The lead–acid cell in the experiment above has a voltage of 2 volts when it is charged. Car batteries usually have 6 of these cells connected together to make a battery with a voltage of 12 volts.

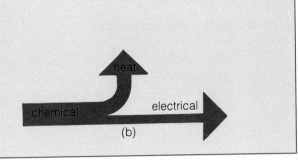

A battery of six lead/acid cells

5 Copy this diagram of 6 cells and draw in wires connecting them together to make a battery of 12 volts.

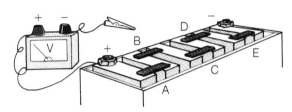

6 The metal strips connecting the cells in the diagram above are labelled A,B,C,D,E. What voltage reading would you get when you connect a voltmeter between the (+) terminal and each strip? Each cell has a voltage of 2 volts.

Strip	Voltage
A	
B	
C	
D	
E	

Recharging a lead–acid battery

EXPERIMENT. Using a battery charger

A battery charger has two leads marked (+) and (−). Connect these leads to the battery (+ to +, − to −) and adjust the charging current to about 3 A. If a larger current than this is used the plates of the battery may be damaged. A battery can be recharged with energy from the mains electricity supply but the recharging has to be done slowly over a long period. The battery is fully charged when the density of the sulphuric acid is 1.2 g/cm^3 (use a battery hydrometer to check this) or when bubbles of gas appear in the acid.

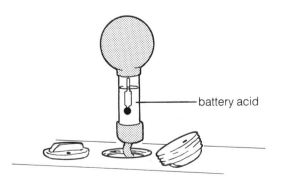

Using a battery hydrometer to find if the battery is charged

Other checks on the battery

While the battery is charging check the level of the acid in each cell. Use distilled (pure) water to 'top up' the acid level to just above the plates. Keep the terminals clean and lightly greased.

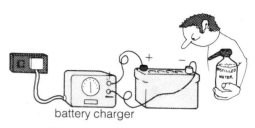

Charging a lead/acid battery

Why do cars need batteries?

Cars with petrol engines have a battery to work their many electrical parts. Current is needed by three circuits:

(a) The starter motor circuit. The electric starter motor uses a large current to start the engine.
(b) The ignition system. This circuit uses current to produce the spark in the cylinders that keeps the engine running.
(c) The lighting circuit. The lights are vital for signalling and good visibility. A lead–acid battery is used in cars because it can provide large currents and can be recharged. A dynamo (or alternator), driven by the engine, recharges the battery automatically and keeps it fully charged.

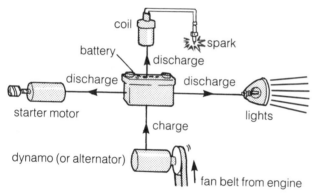

The way a battery is normally used in a car

> **7** What will happen to the battery in a car if the fan belt breaks while the car is being driven along? Will the engine stop?

Other voltage supplies

Batteries are an expensive way of providing electrical energy. A cheaper way is to use a 'low-voltage box' that draws its energy from the mains electricity supply. Low-voltage boxes can usually provide different voltages and have a control for changing the voltage they give.

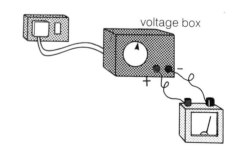

EXPERIMENT

Use a voltmeter to check the voltages from a low-voltage box.

Choose a voltmeter that can read up to the maximum voltage of the box. Plug it directly into the box terminals (red to red, black to black) and measure the reading for each voltage setting. Put your results into a neat table.

8 What are the readings shown on these voltmeters.

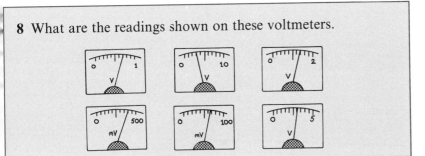

Electric circuits

EXPERIMENT. The simplest circuit

Use a low-voltage box and two pieces of wire to light a lamp. First find the working voltage of the lamp (usually written on the glass) and adjust the voltage box to give the same value. Then connect the wires so that the lamp lights.

When each terminal of the voltage box is connected to the lamp by pieces of metal or wire, the lamp lights. These connections make a closed metal path around which the electricity can flow. This is an electric circuit.

Symbols

Simplified pictures (symbols) are used to draw batteries, lamps and other electrical components. Here are some examples.

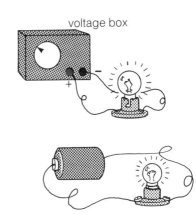

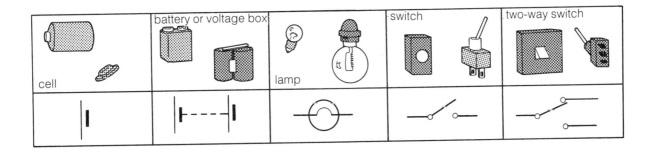

An electric circuit can also be drawn simply as a circuit diagram.

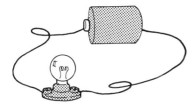
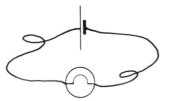
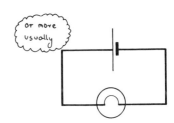

The twists and turns of the wire are usually straightened out. It makes no difference to the flow of electricity if the wires are twisted.

Circuit testers

Here are two circuit building projects.

1 See if you can make this old game. It needs cool nerves and a steady hand. You have to move a ring along a twisty wire without making contact and lighting the lamp.

Use stiff wire for the ring and the twisty section, and thinner copper wire for the rest of the circuit. Check the lamp lights when the ring touches the twisty wire and then see if you can negotiate the bends without making contact.

The game is a simple circuit with a gap between the ring and the twisty wire. As soon as the gap is closed, the circuit is completed and the lamp lights.

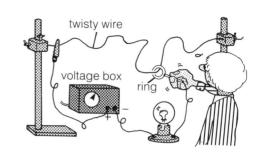

2. Odd ends. This project is an example of the use of electricity to find whether wires are connected together.

Make some U-shaped pieces of bare wire and a few straight pieces. Bury them in plasticine so that the uprights all look the same. Now ask a friend to use a lamp circuit to find which ends are connected together.

You could use the same circuit to test if fuses have broken inside.

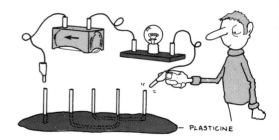

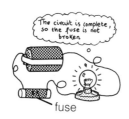

Switches

If there is a break in a circuit at any point, the electric current will not flow. Most circuits contain a switch. A switch is a quick and convenient way of breaking the circuit and switching off the electric current.

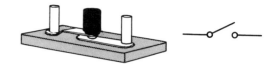

A 'switch' that stops a flow of people

EXPERIMENT. Sending messages at the speed of light

Build this circuit with a voltage box, a press switch and a lamp. Test if it works and then see if you can flash a message across the room to a friend. Use the Morse code (or invent your own) by giving short and long flashes for the dots and dashes.

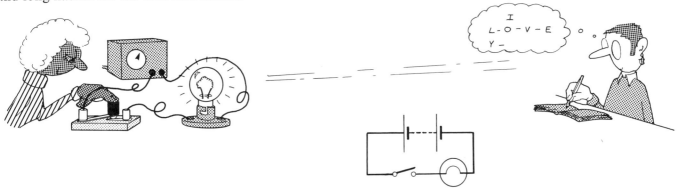

The Morse code

A	·—	J	·———	S	···
B	—···	K	—·—	T	—
C	—·—·	L	·—··	U	··—
D	—··	M	——	V	···—
E	·	N	—·	W	·——
F	··—·	O	———	X	—··—
G	——·	P	·——·	Y	—·——
H	····	Q	——·—	Z	——··
I	··	R	·—·		

Although the message travels at the speed of light, decoding it may take you a little longer.

Truth tables

A 'truth table' is a way of showing how the switches in a circuit work the lamps:
'0' means the lamp is OFF.
'1' means the lamp is ON.
The truth table for this circuit is easy to work out.

TRUTH TABLE

Switch	Lamp
OFF	0
ON	1

Two-switch circuits

Sometimes a circuit contains two switches for a special purpose. Build these circuits and make up a truth table for each one.

EXPERIMENT. A belt-and-braces safety circuit

This circuit is easier to switch OFF than ON.

Large workshop machines are often controlled by two switches for safety reasons. To start the machine both switches must be turned on, but to stop it, only one of the switches has to be turned off.

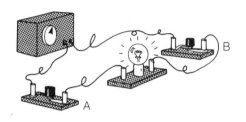
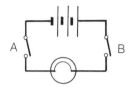

EXPERIMENT. A front-door, back-door circuit

Notice that the lamp can be switched on by either switch. The electric current can go anticlockwise through A, or clockwise through B, or both. But if switch A has put the lamp ON, only switch A can switch it OFF. With this circuit a switch at the front door and another at the back door can turn on the same lamp or bell.

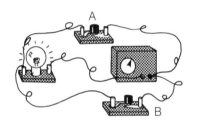
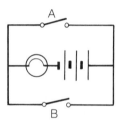

9 These two truth tables are for the two circuits above. Which truth table fits which circuit?

TRUTH TABLE

Switches		Lamp
A	B	
OFF	OFF	0
OFF	ON	1
ON	OFF	1
ON	ON	1

TRUTH TABLE

Switches		Lamp
A	B	
OFF	OFF	0
OFF	ON	0
ON	OFF	0
ON	ON	1

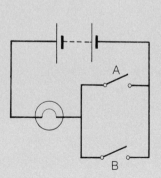

10 This circuit is the same (electrically) as one of the circuits above. Find which it is by working out its truth table and comparing with the ones in question 9.

Two-way switches

This type of switch can be used to direct the electric current along one of two paths. The plank at this part of the marathon race is a two-way switch. It can direct runners to the top island, or if swung over, to the bottom island.

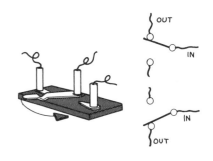

EXPERIMENT. Circuits using two-way switches

Build these circuits and trace the path of the electric current with the switches in each of their positions.

 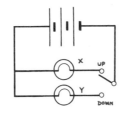

TRUTH TABLES

Switch	Lamps	
	X	Y
UP		
DOWN		

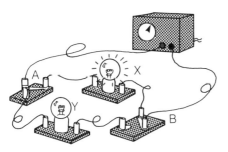 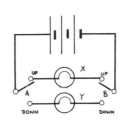

Switches		Lamps	
A	B	X	Y
UP	UP		
UP	DOWN		
DOWN	UP		
DOWN	DOWN		

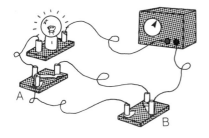 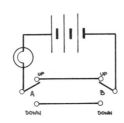

Switches		Lamp
A	B	
UP	UP	
UP	DOWN	
DOWN	UP	
DOWN	DOWN	

11 Copy and complete the truth tables for each circuit.

The last circuit is useful because the lamp can be put ON by one switch and OFF by the other. A lamp above the stairs or in a long corridor is often connected to two switches in this way.

Lighting two lamps at once

Here is one way of lighting two lamps from one voltage supply. The lamps are connected in line so that the electric current passes first through one lamp and then the other. The lamps are connected in series.

Two lamps in series

Build the circuit and notice that there are two problems with series connection:
(a) the lamps are not fully lit (they are sharing the voltage);
(b) if one lamp is removed (or breaks) the other goes out.

At this part of the marathon race, the runners have to cross two bridges that are in series. The runners have to cross one bridge first and then the other, and all the runners that cross the first bridge cross the second. The narrow bridges slow down the whole race just as the lamps slow down the flow of electrons from the battery.

If one of the bridges is blown up, the race stops and no runners can cross the other bridge. This is like removing a lamp from the circuit and stopping the electricity flow.

12 Do you think the lights in your house are connected in series? Explain your thinking.

13 Which of these circuit diagrams show lamps connected in series?

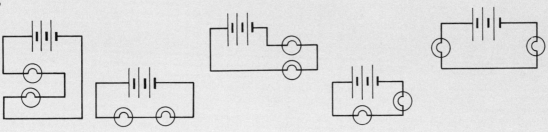

Another way to light the lamps uses an extra wire as shown here. The lamps are connected in parallel and each lamp has its own direct links with the battery. In a parallel circuit: both lamps are fully lit and if one lamp is removed, the other keeps shining.

Build the circuit and check this for yourself.

Notice in this case that the electric current divides when it reaches the signpost. Part of it goes through one lamp and the rest through the other.

Two lamps in parallel

At this part of the marathon route, the runners have come to two bridges in parallel across a river. Each runner has to choose which bridge to cross and those that cross one bridge do not cross the other. It is easier for the runners to cross two bridges in parallel than two bridges in series and the pace of the race is faster.

If one bridge collapses the race can continue across the other bridge, but the pace will be slower than it was before.

14 House lights are wired in parallel. Give two reasons why this is better than wiring them in series.

15 (a) In which circuit would the lamps be brightest?
(b) In which circuit would the lamps be dimmest?
(c) Write down the circuits in order of lamp brightness, brightest first.

22 Electric resistance

It is clear that electric current can flow through copper wires and the filaments of lamps. Are there other materials through which electric current can pass?

EXPERIMENT. Sorting out conductors from insulators

You will need to build a test circuit for this experiment.

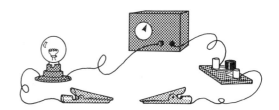

Push crocodile clips onto the ends of two leads and build them into a circuit with a voltage box, switch and lamp. Different materials can be used to fill the gap in the circuit by connecting them between the clips. The lamp will then show whether electric current can pass.

Get together an interesting collection of everyday materials. Use your test circuit to put them into two sets: those that allow electricity to pass (conductors) and those that do not (insulators).

EXPERIMENT. Sorting out the really good conductors

You will have noticed that electric current passes very easily through some of the conductors and the lamp in the circuit shines with full brightness. Test the conductors again and separate the really good conductors from the rest.

The remaining materials do conduct electricity, but the electric current has some difficulty in getting through. The materials have a resistance that reduces the flow of electricity in the circuit. (The very good conductors also have resistance, but it is so small that it makes little difference to the brightness of the lamp.)

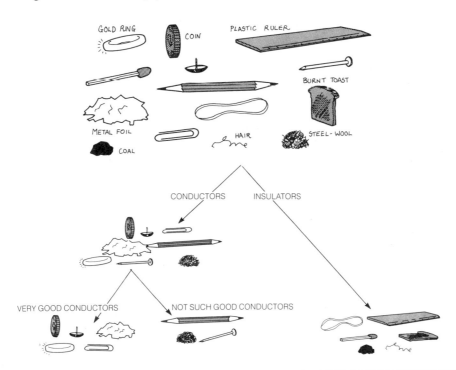

1 Which group of materials contains:
   (a) insulators,
   (b) conductors,
   (c) very good conductors.

A	B	C
Rubber band	Gold ring	Pencil lead
Plastic ruler	Aluminium foil	Wire wool
Human hair	Drawing pin	

Making it difficult on purpose – resistance wire

The resistance of materials can be put to good use, and some wires are specially made to have resistance. Eureka and Nichrome are alloys that are used to make resistance wire. Eureka is made by mixing molten copper and nickel, and Nichrome by mixing nickel and chromium. A wire with resistance is called a resistor and its circuit symbol is —☐—

EXPERIMENT. Resistance depends on length and thickness

Make a coil of resistance wire and clip it into your test circuit. Make the coil shorter by pulling one end and find out if the lamp gets brighter (less resistance) or dimmer (more resistance). Straighten the coil and double up the wire to make it thicker. Find out now whether thick wire has more or less resistance than thin wire.

You will probably find that the resistance of wire:
   increases if it is longer,
   decreases if it is shorter,
   increases if it is thinner,
   decreases if it is thicker.

2 Copy and complete this table about resistance wire.

Shape of the wire	Resistance (high/low)	Lamp (dim/bright)	Current (high/low)
long			
short			
thick			
thin			

3 These shapes are all made from pieces of a soft material that conducts electricity. Which shape will have:
(a) the greatest resistance,
(b) the smallest resistance?
Put all the shapes in order of resistance, greatest first.

4 In which box would you place each of the following phrases:
(a) best conductor,
(b) best resistor,
(c) worst resistor,
(d) worst conductor?

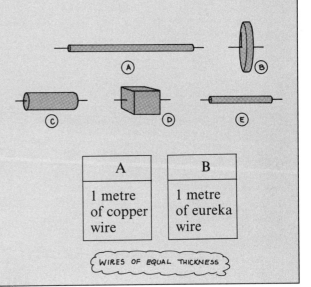

Variable resistors

EXPERIMENT. Current control

Lay 1 m of resistance wire in a large loop on the table and connect it into this circuit. Move the clip B along the wire and notice that as it moves away from A, the lamp gets dimmer. The wire and clip make a simple 'dimmer control' for the lamp.

If B is moved to include a larger length of wire, the resistance in the circuit goes up and the current goes down. The wire and sliding clip are a variable resistor that can control the current in the circuit.

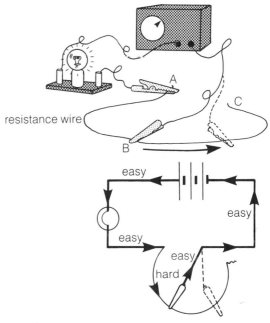

209

A neater resistor

Long lengths of resistance wire can get into a tangle and are awkward to use. Neater resistors can be made from other conducting materials such as carbon powder, sometimes mixed with clay. These small, fixed-value resistors form part of nearly every electronic circuit. Their resistance value is marked on them as a number or by using a code of coloured rings.

EXPERIMENT. Another way of making smaller resistors

Instead of laying out the resistance wire in a large loop, wind it on a pencil and make it into a neat coil. Then use it as a variable resistor by running the crocodile clip across the loops of the coil.

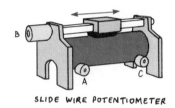

The potentiometer

A resistor that can be varied is so useful that it is made as a special component and given the name **potentiometer**.

One type has a coil of resistance wire, with terminals (A and C) attached to each end. A slider, attached to a third terminal (B), makes contact with the coil and can be slid from one end to the other.

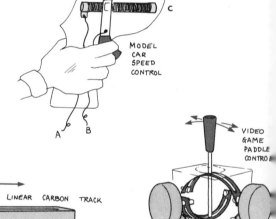

SLIDE WIRE POTENTIOMETER

Practical pots

You probably use a potentiometer ('pot' for short) at home every day. The volume, tone and balance controls on a music centre; the brightness, contrast and colour controls on a television, are all likely to be potentiometers. By changing resistance or voltage, these knobs give us control over electronic machines.

Here are some examples of the many types of potentiometer in common use. Notice they all have a sliding part and three terminals.

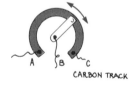

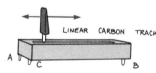
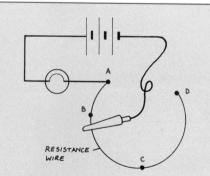

5 Where should the clip be on the loop of resistance wire to make:
 (a) the lamp as bright as possible,
 (b) the resistance as high as possible,
 (c) the current as small as possible?

6 These circuits contain a lamp and control potentiometer. For each one describe what happens to the brightness of the lamp as the slider is moved from the left-hand end to the right-hand end.

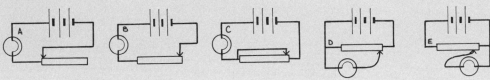

Extra special resistors

EXPERIMENT. What does a diode do?

A **diode** is a widely used component that is made from semiconductor materials. It has a most useful property.

Build these two circuits and find out what this property is.

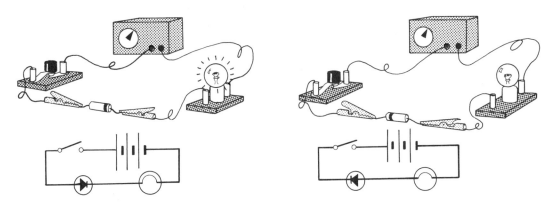

7 Two students did the experiment. One said 'A diode will let electric current pass through in one direction but not in the other'. The other said 'A diode has a low resistance in one direction but a high resistance in the opposite direction'. Explain how they are both right.

The circuit symbol for a diode contains an arrow. Current can flow in the direction of the arrow (towards the line or dot on the component) but not in the opposite direction.

A one-way door

8 In which of these circuits will the diode allow current to pass and light the lamp?

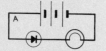 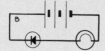  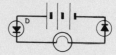

The light-emitting diode

Light-emitting diodes (LEDs) are often used – instead of filament lamps – as indicator lamps on electronic equipment. They are cheap, robust and have a long life. They can be made in different colours and need only a small current to light up brightly. They should always be used with a protective resistance to limit the current they pass.

Experiment with an LED

Clip a light-emitting diode into a circuit with a lamp and a switch. Experiment and see if you agree with the following observations:
(a) an LED only allows current to pass in one direction;
(b) an LED looks brighter when viewed end on than from the side;
(c) the current that fully lights an LED only produces a faint glimmer of light from the lamp.

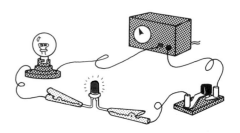

Electric heating

EXPERIMENT. Passing large currents through wires

Build the circuit shown and pass a current through a thin piece of bare copper wire. Increase the current by moving the potentiometer slider and watch what happens to the wire.

Replace the wire with a piece of plastic-covered copper wire, the sort used for connecting circuits. What happens when a large current passes through this wire? Then use lengths of tin wire (or solder), iron wire and Nichrome wire of about the same thickness. Watch what happens with each material and tick the observation if you find it correct.

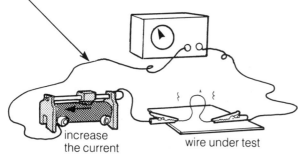

Results

Material	Observations
Thin copper wire	The wire takes a large current before it gets red hot and breaks.
Plastic-covered wire	The plastic melts.
Tin wire	The tin melts at a low temperature. Tin wire is soft and breaks easily.
Iron wire ...	... was able to stay red hot for a while but eventually burnt and broke.
Nichrome ...	... could stay red hot for a long time without damage.
A tungsten lamp filament ...	... could be made white hot without breaking.

9 Using these observations try and answer the following questions about electric heating.
  (a) Give a reason why heating coils of electric fires, irons and kettles are made from Nichrome wire.

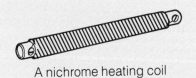
A nichrome heating coil

  (b) Why is tungsten a good material for the filaments of lamps?

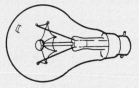

  (c) Why is most of the air in filament lamps replaced by a gas that does not support burning?

A lamp with a tungsten filament

  (d) Tin wire used to be used as fuse wire but has been replaced by copper wire coated with tin. Give a reason why copper is better than tin for fuses.

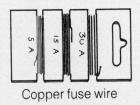
Copper fuse wire

10 Electric heating is quick, easy and clean. Explain the meaning of this cartoon.

Fuses

A fuse is a piece of thin wire that is built into a circuit to protect it from large currents. If, for some reason, the current in the circuit rises above its normal value, the fuse melts and breaks the circuit. Without the fuse, the large current could melt the insulation of the connecting wires, damage components and even cause fires.

A weak link

EXPERIMENT. A home-made fuse

You can make a fuse for a small lamp from a short length of very fine copper wire. Grip the wire between two crocodile clips and include it in the lamp, battery and switch circuit. When you switch on, the lamp should light and the current should pass through the fuse without its wire melting.

Now find out what happens when you 'short out' the lamp. You do this by joining the lamp terminals with a piece of copper wire or a paper clip. You should see the lamp go out and the fuse wire melt.

The fuse allows the lamp to light but as soon as the current rises above its normal value the fuse wire melts and the circuit is broken.

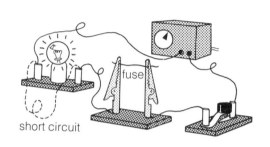

Beware of shorts

The wire connected across the lamp is called a short circuit (or 'short' for short). It should really be called an 'easy circuit' because it provides an easier route for the current than through the thin filament of the lamp. This is why the lamp goes out. Also because the short circuit has a very low resistance, the current in the circuit rises and melts the fuse.

'Shorts' are always bad news and can be dangerous – they allow large currents to flow from the battery. This will quickly drain the battery, overheat the wires, melt fuses and cause damage to components. So check every circuit for these damaging low-resistance links.

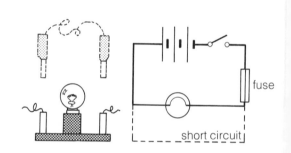

A short circuit in the race

11 When a large current flows through a fuse wire, which box gives the correct sequence of events:

A. The wire melts.
   The wire gets hot.
   The current is cut off.

B. The wire gets hot.
   The wire melts.
   The current is cut off.

C. The wire gets hot.
   The current is cut off.
   The wire melts.

D. The current is cut off.
   The wire gets hot.
   The wire melts.

E. The wire melts.
   The current is cut off.
   The wire gets hot.

12 Which of these circuits contains a 'short circuit'?

   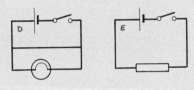

13 Where would you put a fuse in the circuit on the right to protect the whole circuit from too much current?

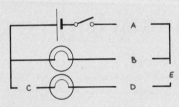

14 Which is the correct symbol for a fuse?

Measuring electric current

The brightness of a lamp is one way of measuring electric current. A better way is to use an **ammeter**. This instrument measures current in units called **amperes**. An ammeter has its terminals marked + and −. To give correct readings it must be connected the right way round. The + terminal of the cell should go to the + terminal of the ammeter and there must be a lamp or other component in series with the ammeter.

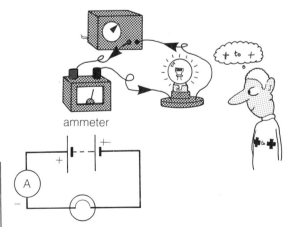

Connecting an ammeter to measure current in a circuit

15 An ammeter can be connected in two positions between the lamp and the cell. One position is shown. Draw a circuit diagram showing the ammeter in the other position. Will it give the same reading?

The direction of an electric current

The current, we say, flows from the + terminal of the cell through the ammeter and lamp to the − terminal of the cell. This is the 'conventional' direction of the current flow. (The electrons flow the other way.)

Measuring resistance

EXPERIMENT

Use an ammeter to measure the current that passes through some common conductors when they are connected to a 12 V power supply. Although the voltage is the same for each conductor, the current that flows is different. This shows that each conductor has a different resistance.

Record your results in a table like the one in question 16. The readings will not be simple numbers so use a calculator to work out the voltage/current column.

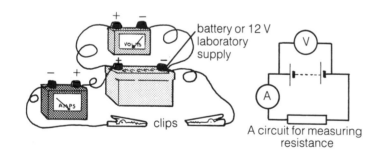

A circuit for measuring resistance

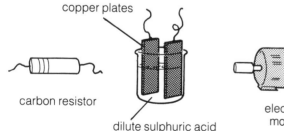

carbon resistor

copper plates

dilute sulphuric acid

electric motor

electric heater

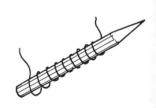

torch bulb

1 metre of resistance wire

16 The current and voltage readings in the table below were taken from an experiment like this.

Device	Carbon resistor	Dilute sulphuric acid	Torch bulb	Motor	Heater	Resistance wire
Voltage (V)	12	12	12	12	12	12
Current (A)	$\frac{1}{4}$	2	$\frac{1}{2}$	4	3	1
Voltage/Current						
Forms of energy produced						

(a) Which device has (i) the greatest resistance, (ii) the smallest resistance?
(b) Copy the table and calculate voltage ÷ current for each device (e.g. for the carbon resistor: $12 \div \frac{1}{4} = 48$).

You will notice that the value of voltage ÷ current is largest for the conductors with the largest resistance (when the current is smallest). For this reason voltage ÷ current is called the **resistance** of the device.

$$\text{resistance} = \frac{\text{voltage}}{\text{current}}$$

$$\text{or } R = \frac{V}{I} \quad \text{using the usual symbols.}$$

Units

If 1 volt drives a current of 1 ampere through a conductor, then that conductor has a resistance of 1 ohm (symbol Ω). The unit of resistance is the **ohm**.

17 Write down the conductors in question 16 in order of their resistances, the lowest one first.

18 Each device changes electrical energy into different forms. Write the forms of energy produced by each device in the last row of boxes in question 16.

EXPERIMENT. Measuring resistance with a multimeter

A multimeter can measure voltage, current or resistance. It has a dial or buttons that allow you to choose the one you want to measure.

Adjust the meter to a resistance range and use it to measure the resistance of the conductors in the previous experiment.

No voltage supply is needed because the meter has its own battery. It also does the calculation for you.

Do your two sets of results agree?

EXPERIMENT. Does resistance change with voltage?

This important experiment measures the resistance of a coil of Nichrome wire at different voltages.

Make a coil from a long length of fine Nichrome wire. Clip it into the circuit shown.

Turn the voltage down to its smallest value and take readings of the current and voltage. Then increase the voltage in steps, taking voltage and current readings as you go. (Between readings you should switch off the supply and allow the coil to cool.)

Put the results into a table and calculate the resistance of the coil for each voltage setting.

Does the resistance change as the voltage goes up?

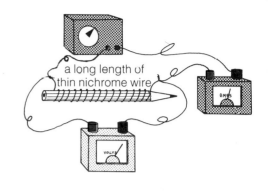

19 These simplified readings were obtained from an experiment like this.
   (a) Calculate the resistance (voltage ÷ current) for each voltage.
   (b) Which of these sentences are correct according to the results?
      (i) the current stays the same whatever voltage is used;
      (ii) the current goes down as the voltage goes up;
      (iii) the current rises as the voltage rises;
      (iv) doubling the voltage, doubles the current.
      (v) voltage/current always gives the same number (a constant).

Voltage used (V)	Current (A)	Voltage / Current (Ω)
0	0.0	
2	0.1	
4	0.2	
6	0.3	
8	0.4	
10	0.5	

Ohm's law

For the resistance in question 19, dividing the voltage by the current comes to 20 each time. Are your results constant? (Don't forget that it is impossible to do a perfect experiment and your results will include some errors.)

Another way of describing the findings is to say that the current is proportional to the voltage. That is, the current doubles if the voltage doubles etc.

You can test for direct proportion between voltage and current by plotting a graph of them. A straight line through the origin shows that they are proportional.

Careful experiments have found that for many conductors: 'the current through a conductor is proportional to the voltage difference across it, provided its temperature does not change'. This is known as Ohm's law.

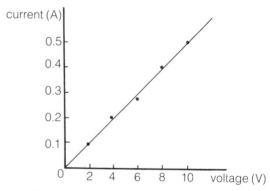

Current and voltage are proportional

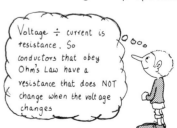

20 Describe how you would find out whether a conducting material obeys Ohm's law.

Do all resistors follow Ohm's law?

A student heard that not all circuit components with resistance obey Ohm's law and so set out to check this. She built a basic test circuit using a variable voltage supply, voltmeter and ammeter that could be used for all components. Then for each component she measured the p.d. and current for a range of voltage settings. (The ammeter had a number of ranges that could be selected to suit each component.)

Her readings are given below. If possible you should do the experiment and get your own results.

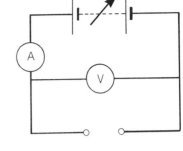

Lamp filament (a coil of thin tungsten wire).

Voltage (V)	0	0.5	1.0	2.0	3.0	4.0	5.0	6.0
Current (A)	0	0.11	0.14	0.19	0.23	0.27	0.31	0.34
Resistance (Ω)								

Semiconductor diode (a one-way current gate).

Voltage (V)	0	0.1	0.2	0.3	0.4	0.5	0.6	0.7
Current (mA)	0	0	0	0	0	1.5	73	250
Resistance (Ω)								

Light-emitting diode (a solid state diode that lights as it conducts).

Voltage (V)	0	1	1.5	2	3	4	4.5	5
Current (mA)	0	0	0.2	6.9	42	86	118	163
Resistance (Ω)								

Note how rapidly the current rises. An LED should normally have a resistance in series to protect it from large currents.

Thermistor (a component whose resistance changes widely with temperature).

Voltage (V)	0	0.5	1	1.5	2	3	4	5
Current (mA)	0	0.7	1.4	2.1	2.8	4.2	5.6	7.0
Resistance (Ω)								

Copper II sulphate solution with copper plates (a liquid conductor).

Voltage (V)	0	0.5	1	2	3	4	4.5	5
Current (A)	0	0.07	0.15	0.30	0.44	0.59	0.67	0.74
Resistance (Ω)								

Light-dependent resistor (a photocell whose resistance changes with light level).

Voltage (V)	0	0.5	1	2	3	4	4.5	5
Current (mA)	0	10	21	43	65	88	98	110
Resistance (Ω)								

---

21 For each of the components:
  (a) calculate the resistance for each voltage;
  (b) plot a graph of current against voltage;
  (c) decide whether the component follows Ohm's law or not.
  (Why is it impossible to decide whether the lamp filament follows Ohm's law?)

Two more extra special resistors

The thermistor and light-dependent resistor (LDR) are examples of components that can be used as sensors in electronic circuits. Thermistors can detect changes in temperature and LDRs respond to changes in light level.

thermistor

LDR

◐ EXPERIMENT. Heating a thermistor

Connect a thermistor into this circuit and use the meters to measure its resistance at different temperatures. Start with boiling water in the beaker and get resistance values as the water cools. Use a thermometer to measure the temperature of the water, taking care to stir well before each reading.

Plot the readings on a graph to show how the resistance of a thermistor varies with temperature.

Use the same apparatus to find how the resistance of a coil of fine iron wire changes with temperature. Plot a graph for the iron and compare it with the thermistor results.

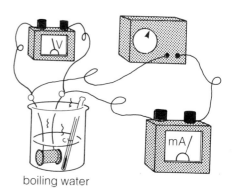
boiling water

22 Here are results from an experiment with a thermistor.

Temperature (°C)	23	37	47	57	65	76	85	90	100
Voltage (V)	5	5	5	5	5	5	5	5	5
Current (mA)	19	24	31	37	45	56	67	73	87
Resistance (Ω)									

(a) Calculate the resistance for each temperature and plot a resistance/temperature graph for the thermistor.
(b) Describe how its resistance changes with temperature.
(c) When the thermistor was dipped in a cup of tea its resistance was found to be 100 Ω. Use your graph to estimate the temperature of the tea.

23 The current readings for a coil of iron wire, for the same voltages as in question 22 are:

Current (A)	0.50	0.47	0.42	0.36	0.34	0.30
Temperature (°C)	0	10	30	50	70	100

Plot a graph that shows how the resistance of iron wire changes with temperature.

24 This network is part of a TV circuit where a thermistor is used to protect the set from overheating. Explain why an unusual rise in temperature will cause the fuse to melt.

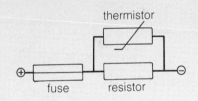

EXPERIMENT. Light on an LDR

Cover a light-dependent resistor with a cap made from black card. Make a single pin-hole in the cap and fix a lamp above the hole. Measure the LDR's resistance for this light level. Now make a second equal hole to double the light level and measure the resistance again. Continue to make holes and measure resistance until the cap is covered in holes.

To show how the resistance of the LDR varies with light level, plot a graph of your resistance values against the number of holes.

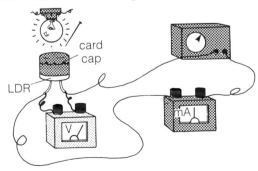

25 Here are results from such an experiment.

Number of holes	1	2	3	4	5	6	7	8
Voltage (V)	5	5	5	5	5	5	5	5
Current (mA)	2	2.8	3.3	3.7	3.9	4.1	4.3	4.4
Resistance (Ω)								

(a) Plot a graph of resistance against number of holes.
(b) Describe how the resistance of an LDR varies with light level.
(c) Sketch a second graph to show the curve you would expect to get if the lamp were fixed closer to the cap.

Adding resistors together

Two resistors can either be joined end to end (in series) or side by side (in parallel).

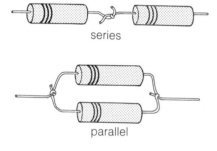

This diagram shows all the networks that can be made from three equal resistors.

26 Which circuits contain resistors that are:
 (a) in series only,
 (b) in parallel only,
 (c) in series and parallel?

27 These are eight different ways that a fourth resistor can be added to circuits D, E, F, G. Draw all the possible networks of four equal resistors.

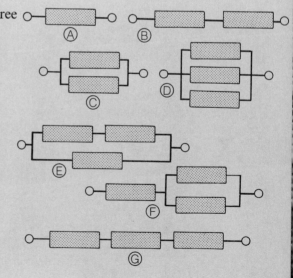

Finding the total resistance

Series
When resistors are joined together in series their resistances add up.

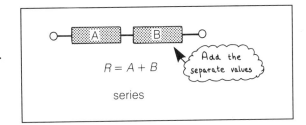
$R = A + B$

series — Add the separate values

Parallel
When resistors are joined in parallel the total resistance goes down. (This is because the current can pass by two routes.) The formula for working out the total resistance is shown in the 'parallel' box.

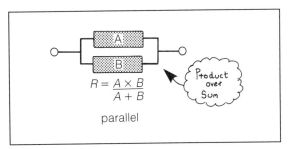
$R = \dfrac{A \times B}{A + B}$

parallel — Product over Sum

For example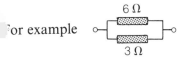

has a resistance of $\dfrac{6 \times 3}{6+3} =$ 2 Ω.

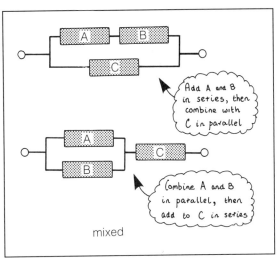

Add A and B in series, then combine with C in parallel

Combine A and B in parallel, then add to C in series

mixed

28 Calculate the total resistance of the networks on p. 222. Each resistor is 6 Ω. Put the networks in resistance order.

29 Calculate the total resistance of these networks of four equal resistors. Each resistor is 8 Ω. Put them in resistance order.

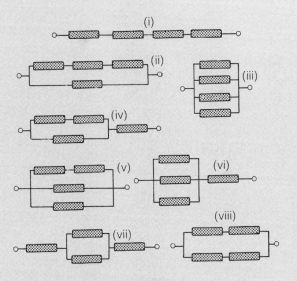

30 You have three resistors with values 6, 3 and 2 Ω. Sketch all the networks you can make with them, either singly or joined together. (There are 14 in all.) Calculate the total resistance of each network.

Current at a junction

At every branch in a circuit the total current that flows in equals the total current that flows out. No current can be 'lost' at a junction nor can more current leave than enters.

31 What would happen at a junction if more current flowed in than flowed out.

Most goes the easy way

The current divides according to the resistance of each path. If the branches have equal resistances, the same current will flow through each. But if the branches have different resistances, the current will divide unequally and the larger fraction will go through the branch with the smaller resistance.

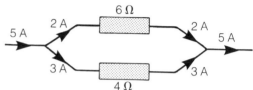

Calculating the fractions

Add the two resistances and divide the current into that number of parts – 'thirds' in the first example. Then $\frac{2}{3}$ will go through the 1 Ω and $\frac{1}{3}$ through the 2 Ω resistor.

Here are some other examples. Note that 'current × resistance' is the same for each resistor since the p.d. is the same across each branch.

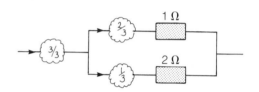

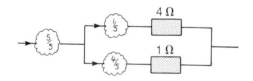

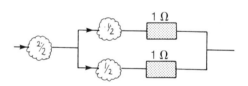

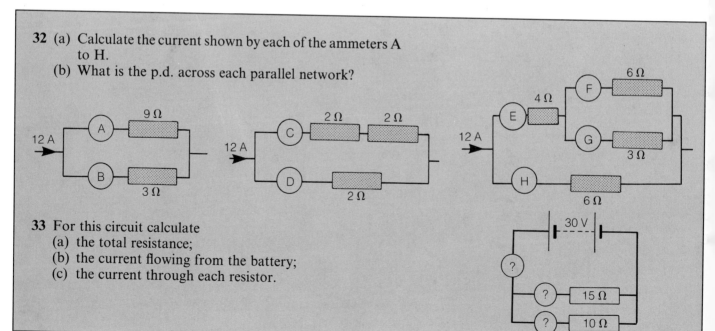

32 (a) Calculate the current shown by each of the ammeters A to H.
 (b) What is the p.d. across each parallel network?

33 For this circuit calculate
 (a) the total resistance;
 (b) the current flowing from the battery;
 (c) the current through each resistor.

Dividing up a 'voltage'

The voltage of a battery or voltage box can be divided by putting two resistors in series with it.

Build these circuits and you will see how it works.

Two equal resistors divide the voltage into two equal parts. Unequal resistors divide the voltage into unequal parts with the largest voltage across the largest resistor.

The fraction of the voltage across a resistor A is $A/(A+B)$ and across B is $B/(A+B)$.

If one of the resistors is variable then the voltage across it will vary as its resistance is changed.

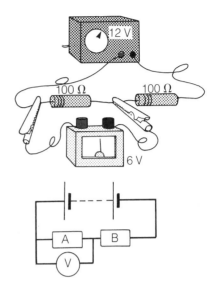

Voltage control

Here are three ways of controlling the voltages in a circuit from outside. They all have practical applications in electronic control. Build each circuit and see for yourself how the voltage in the circuit responds to outside control.

By hand

This circuit shows how a potentiometer (p. 210) can be wired up to give a variable voltage. As the slider is moved from one end to the other, the voltage changes continuously from zero to a maximum. Hand control like this is used to set the volume, balance and tone levels of an amplifier for example.

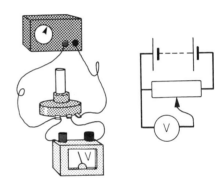

By light

If one of the resistors is an LDR (p. 221), then the voltage across it will change with the light level. In the dark its resistance is high and the voltage across it is high too. In bright light, its resistance is low and the voltage across is also low.

Light sensing circuits can use LDRs in this way (p. 287).

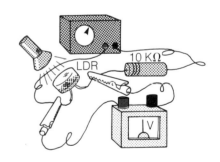

By temperature

Thermistors have a resistance that changes with temperature (p. 220). In this circuit the voltage across the thermistor is high when it is cold (high resistance) and low when it is hot (low resistance).

Electronic thermostats can use thermistors in this way (p. 291).

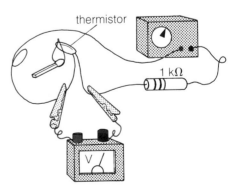

34 The LDR in this circuit has a resistance of 4400 Ω in the dark and 80 Ω in the daylight. Calculate the values of the current and p.d. for this table.

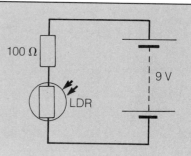

	Dark	Daylight
Current through the LDR		
p.d. across the LDR		

A summary of circuit quantities

To illustrate the difficult ideas of electromotive force (e.m.f.), potential difference (p.d.), charge and resistance consider this circuit.

The battery has an e.m.f. of 12 volts and an internal resistance of 1 Ω. It drives a current of 3 A through itself and an external resistance of 3 Ω. The p.d. across the 3 Ω resistor is 9 volts (from $V = IR$).

An e.m.f. of 12 volts means that each coulomb is given 12 joules of energy as it passes through the battery. 3 joules of that energy are used to get through the battery's own resistance.

A p.d. of 9 volts means that each coulomb uses 9 joules of energy to get through the 3 Ω resistor. This energy is turned into heat. (The p.d. across a resistor = the energy changed by each coulomb that passes through it.)

A current of 3 A means that 3 coulombs of charge pass any point in the circuit each second.

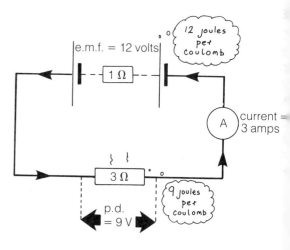

35 Copy this box of definitions and put the following into the correct boxes: coulomb, internal resistance, p.d., e.m.f., current.

	The energy supplied to 1 coulomb of charge as it passes through a battery.
	The energy changed by one coulomb as it moves between two points of a circuit.
	The flow of charge per second.
	The resistance inside a battery.
	The charge that passes when 1 ampere flow for 1 second.

23 Electric power

Measuring electrical energy

We have seen ways that electrical energy can be changed into light, heat, movement and sound. Electrical energy can be measured by a joulemeter. This is rather like a petrol pump that measures the amount of petrol flowing through. The amount of energy is measured in joules.

EXPERIMENT. Using a joulemeter

Connect a lamp to a joulemeter and a 12 volt a.c. supply as shown. Switch on and watch the dials of the joulemeter move round. It is measuring the electrical energy that the lamp is changing into heat and radiation. (The electricity meters in our homes are high-voltage versions of the joulemeter.)

Measure the energy changed in one minute by energy converters such as a headlamp bulb, torch lamp, electric motor, heater. (They must all work on 12 volts a.c.)

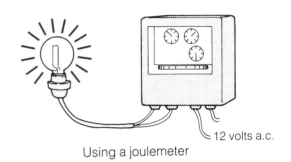

Using a joulemeter

1 The drawings show the dials of a joulemeter before it was switched on and after it was switched off 1 minute later. What are the two readings and how much energy has changed in that minute?

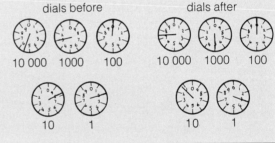

Electric power

You will have noticed that some of the electric energy converters change energy faster than others. For example, a headlamp bulb uses energy more quickly than a torch lamp and produces more light as a result. The headlamp has more power than the torch lamp. Power is the energy changed per second by the device. It can be calculated from the equation

$$\text{power} = \frac{\text{energy changed}}{\text{time taken}}$$

Units
Energy is measured in joules (J).
Time is measured in seconds (s).
Power is measured in watts (W).

2 This table gives the energy changed by each piece of equipment in 1 minute. Copy the table and fill in the power of each device.

	🔌	💡	motor	heater
Energy changed (J)	1200	120	600	3000
Time taken for this (s)	60	60	60	60
Power (W)				
Forms of energy produced				

The power of a device tells us how quickly it changes energy from one form to others.

Examples

A 100 W light bulb: 100 J of electric energy are changed every second into heat and radiation.

A 100 W arm muscle: 100 J of chemical energy are changed every second into gravitational potential energy.

A 100 W burner: 100 J of chemical energy are changed every second into heat and radiation.

3 (a) How much energy will each of these converters change in one second?
  (b) How long will it take each one to change the same energy as the more powerful version can in 1 second.

2 W

10 W

20 W

The electrical power of a device depends on the voltage and current it uses. Power can be calculated from this equation (see p. 231):

$$\text{electric power (W)} = \text{voltage (V)} \times \text{current (A)}$$

$$\boxed{P = VI}$$

EXPERIMENT. Another way of measuring electrical power

Build this circuit by first connecting a heater (or lamp) and ammeter in series with a voltage supply. Then add the voltmeter across the terminals of the heater. Use the meters to measure the voltage across the heater and the current that flows.

Then calculate the power of the heater from the power equation.

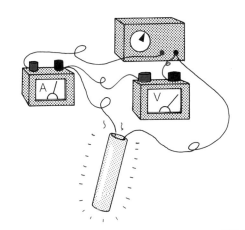

4 Draw the circuit diagram for this experiment. Use these symbols. (The heater is just a resistor.)

5 Calculate the power of the following items. Copy the table and fill in the spaces.

	Tube light	Mains lamp	Electric kettle	One-bar electric fire	Transistor radio
Voltage (V)	240	240	240	240	9
Current (A)	$\frac{1}{6}$	$\frac{1}{2}$	10	4	$\frac{1}{9}$
Power (W)					

6 Voltage and current values can also be used to calculate resistance. Copy this table and work out the answers that go in the spaces.

	Fridge	Iron	Television
Voltage (V)	240	240	240
Current (A)			1
Resistance (Ω)		80	
Power (W)	80		

7 You have calculated the power of 12 electrical devices on the last two pages. Write a list of them in order of the power they use.

Paying for electrical energy

The dials of your electricity meter at home measure electrical energy in kilowatt hours (kW h).

1 kilowatt hour is the amount of energy used by a 1000 W appliance in 1 hour (1000 watts for 1 hour).

Electrical energy is sold in these units of kilowatt hour. Electricity Boards call one kilowatt hour a 'unit'.

One 'unit' of electrical energy is therefore enough to run a:

   1 kW electric fire for 1 hour
   3 kW kettle   for $\frac{1}{3}$ hour (20 minutes)
500 W electric iron for 2 hours
100 W lamp   for 10 hours

The amount of energy used by a:

   3 kW immersion heater in 4 hours is   12 kW h (12 'units')
250 W   colour TV   in 8 hours is   2 kW h
100 W   hair drier   in 1 hour is   0.1 kW h
   2 kW fire   in 10 minutes is 0.33 kW h
                                (change minutes to hours)

Joule and kilowatt hour are both energy units.
   1 kilowatt   is 1000 J/s
   1 hour   is 3600 s
so 1 kilowatt hour is 3600×1000 J

$$1 \text{ kW h} = 3\,600\,000 \text{ J}$$

8 Calculate the quantities that go in the blank boxes.

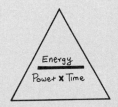

Appliance	Power (W)	Power (kW)	Time used (h)	Energy used (kW h)
Kettle	3000		$\frac{1}{2}$	
Radiator		2		8
Tube light	50		10	
Vacuum cleaner			6	3

Proof of the electric power equation

If the p.d. across a conductor is V volts and the current through it I amps, then its power $P = VI$.

A p.d. of V volts means that each coulomb of charge that passes converts V joules of electrical energy into other forms.

A current of I amps is a flow of I coulombs each second.

So the electrical energy converted each second by the conductor is VI joules/second (or watts). And

$$P = VI$$
power = p.d. × current

9 Use the resistance equation $R = V/I$ to derive these alternative power equations:

$$P = \frac{V^2}{R} \quad \text{(substitute for } I\text{)}$$

$$P = I^2 R \quad \text{(substitute for } V\text{)}$$

10 If a p.d. V drives a current I through a resistor for a time t, show that the total amount of energy converted E is given by:

$$E = VIt$$

11 Calculate the unknowns in this table for the resistors A to D.

	A	B	C	D
p.d., V (V)	12	9	250	
Current, I (mA)		180		5
Resistance, R (Ω)	240			
Power, P (W)			0.1	2

12 A 20 Ω resistor was connected to a joulemeter which supplied it with a constant p.d. of 12 V. The resistor used 100 J of energy in 20 s. For the resistor calculate:
(a) its power,
(b) its current.

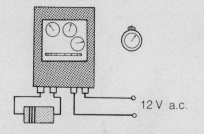

13 A p.d. of 12 V is maintained between the points X and Y of this resistor network. Calculate:
(a) the current entering the network,
(b) the current through each resistor,
(c) the power of each resistor.

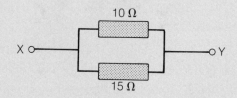

14 Compare the total power used by two 72 Ω resistors when connected first in series and then in parallel to a 12 V supply.

15 A battery that supplies a p.d. of 12 V is connected to a 4 Ω resistor and a variable resistance R. Complete the table by calculating the current and power developed in R for values of R from 1 Ω to 8 Ω.

R (Ω)	1	2	3	4	5	6	7	8
I (A)								
P (W)								

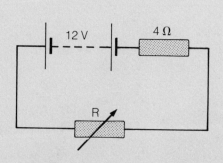

Plot a graph of this power P against R. Use the graph to find the value of R that would draw the most power from the battery.

24 Electricity in the home

Electricity is supplied to your home at a high voltage (220–250 V). To touch any bare wire or metal connected to the mains electricity supply will drive a current through your body that could kill you. Never experiment with the 'mains'. It is highly dangerous for a person without proper training to touch household circuits. Two wires bring the electricity into your home. One wire (live) is at a high voltage and the other (neutral) is at 0 volts. Appliances are connected to the live and neutral wires and the voltage drives current through them.

The electricity meter

The live and neutral wires first pass through the electricity meter. This meter measures the amount of electrical energy used in the home. If the meter has dials, they are read in the same way as the dials on a joulemeter (p. 227).

The consumer unit

The live and neutral wires then pass into the **consumer unit**. The unit contains a switch and a number of fuses. You should know where this switch is because it can switch off all the electricity in the home. The fuses protect the electric circuits that go around the home. These fuses are usually wire (or cartridge) push-in fuses (p. 237). If the fuse wire melts it can be replaced by spare fuse wire of the same value. But first the electricity must be switched off and the fault traced.

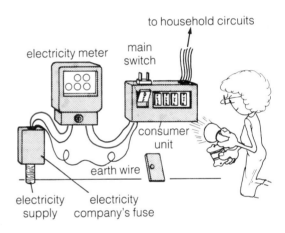

How electricity enters a home

The earth wire

A third wire connects the consumer unit to a plate or pipe in the ground. This is called the earth wire. Its use will become clear later.

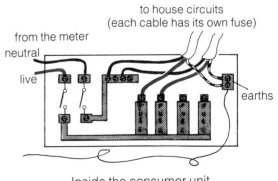

Inside the consumer unit

Electric cables with three wires inside leave the consumer unit and carry the electricity around the house. These cables are usually under the floor or buried in the walls. There are three main types of circuit.

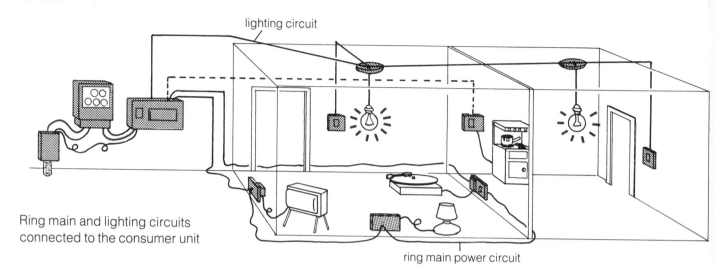

Ring main and lighting circuits connected to the consumer unit

The lighting circuit

The lights in a house are always connected in parallel (p. 205). (What would happen if the lamps were in series and one lamp went out?) The live wire to each lamp passes through a switch. This makes sure the lamp socket is not live when the switch is off. One terminal of the **switch** however is always live (unless the **main** switch is off).

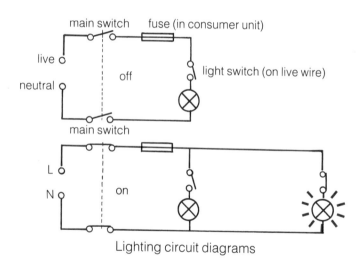

Lighting circuit diagrams

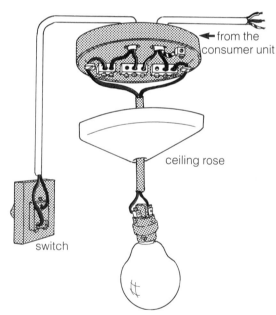

How a lamp and switch are wired up

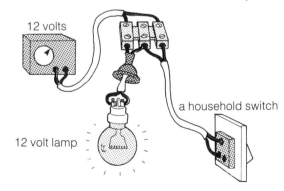

EXPERIMENT

Use two-core cable to connect a lamp and household switch to a battery or low-voltage electricity supply. Follow the diagram. If you have not got a household switch use a simple laboratory switch.

The ring main power circuit

This circuit uses cable with three wires, live, neutral and earth, and connects sockets to the consumer unit. The live and neutral wires are connected to the bottom two holes of each socket. These holes are the ones through which the current enters and leaves. The earth wire is connected to the larger hole at the top. Note that each wire forms a 'ring' starting and finishing at the consumer unit. It takes a little more cable to join the sockets in a ring. The advantage is that current can reach a socket by two routes and each wire has to carry only half the current.

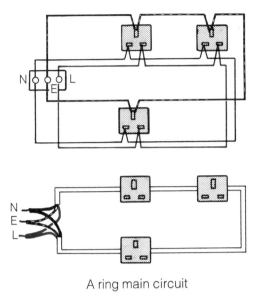

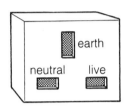

A ring main circuit

Plugging into the ring main

Three-pin plugs are used to connect electrical equipment into the sockets of the ring main. The diagram shows three-core flex connecting a 3-pin plug to an electric iron. The wires from the live and neutral pins of the plug are connected to the heating element of the iron. The current flows through these wires to the heating element. The earth pin of the plug is connected to the metal case of the iron. The earth wire carries no current in normal use.

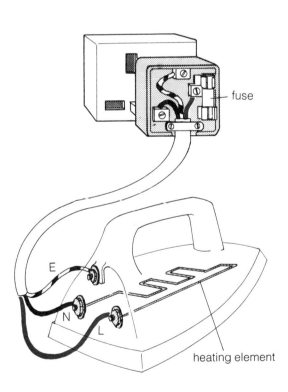

Special circuits

Equipment that uses large currents – electric cookers, water heater, 'instant' showers – are not plugged into the ring main. The device is connected by its own cable to the consumer unit where it has its own fuse.

Earthing

What are the earth pin and earth wire for?

Look at the iron in the picture. It is not earthed. If a loose live wire touches the case, it would be raised to 240 V. If you touched the case, this voltage could drive enough current through your body to kill you. If the case is properly earthed contact with live wire would cause a large current to pass along the earth wire into the ground. The fuse in the plug would melt, cutting off the electricity supply. It is best to earth electrical equipment that is meant to be handled.

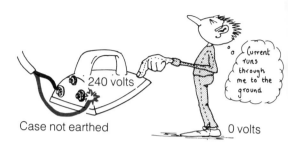

Case not earthed

Equipment such as electric kettles, irons, guitars, cookers, desk lamps, washing machines, vacuum cleaners, refrigerators and electric heaters should be securely connected to the earth. Some equipment that carries the label ▢ is double insulated. This need not be earthed.

Case earthed

1. What would happen if you connected the earth wire from an electric iron to the live pin of a three-pin plug? (This is very dangerous. Never get the wires of a plug mixed up like this.)

2. Why would it be especially dangerous to touch an iron with a live case if you have wet hands?

3. Why are battery radios and cassette recorders not earthed?

4. Describe the route connecting the case of an earthed appliance to the plate in the ground.

Electric jobs you can do for yourself

Fixing a plug to a flexible cable (flex)

Snip the sleeve of the cable in two places. Pull back and trim off 5 cm of sleeve revealing the three wires inside.

Push the flex under the flex grip and screw it down. Cut each wire just long enough to reach its pin and strip off the last centimetre of plastic insulation. Fix each wire to its proper pin (brown to live; blue to neutral; yellow/green to earth).

Turn the wire clockwise round the screw (or push it right into the hole of the connecting block). No bare wire should be visible when the screw has been tightened up.

Replace the back of the plug.

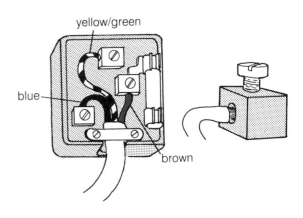

Changing a fuse in a plug
Simply take the back off the plug (having removed it from the socket), lift out the fuse and replace it with one with the correct value (p. 238).

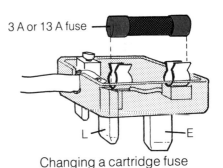

Changing a cartridge fuse

Mending a wire fuse
Have a torch and screwdriver handy, then switch off the main switch. Pull out the fuses and find the one with the broken wire. Take out the broken pieces, clean up any burn marks and fit a new piece of fuse wire of the correct value (usually 30 A, 15 A or 5 A).

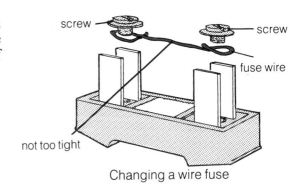

Changing a wire fuse

How to choose a fuse

Fuses sometimes break because they are old but usually a broken fuse means that there is a fault somewhere in the circuit. The fault should be found and put right. Then the fuse must be replaced with one of the correct value.

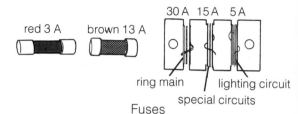
Fuses

How to work out which fuse to use

All electrical equipment should have written on it the power and voltage it uses. For example, an electric kettle may be marked 3000 W, 250 V. This information can be used to calculate the current used and the fuse needed. The electric kettle will use a current of 3000/250 = 12 A. A 13 A fuse will allow a 12 A to pass without melting, but will melt if a fault causes the current to rise to 13 A or more. A 13 A fuse is suitable for this kettle.

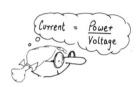

5 Calculate the current and choose the fuse needed for these household electrical items. Normally only 13 A or 3 A fuses are used in plugs. Choose one of these fuses for each item. Copy and complete.

	Electric kettle	One-bar fire	Lamp	Electric iron
Power (W)	3000	1000	125	750
Voltage (V)	250	250	250	250
Current (A)				
Fuse				

Note that vacuum cleaners, television sets, refrigerators and freezers are usually fitted with 13 A fuses, although they often use less than 3 A (there is a surge of current when they first switch on).

6 Make a list of all the electrical equipment you have at home. Against each item write the power it uses and the value of the fuse it should have.

For equipment with power...	up to 720 W	720–3000 W	more than 3000 W
use fuses of...	3 A	13 A	Do not plug into a socket.

Cable and flex ratings

An appliance must be connected to its plug by suitable flex. This must be able to carry the current used by the appliance without getting too hot and melting its insulation (p. 212).

For example, a kettle that uses a current of 12 A must have flex rated at 12 A or more. If thinner flex is used, it could overheat, melt its insulation and 'short' or catch fire.

Safety

Electricity brings many benefits into our homes but it can bring danger too. Apart from fires, electricity can cause serious injury. It is best to be very careful when using electricity at home. Here are some common sense rules to prevent injury to you and your family.

1. Do not plug anything into light sockets except light bulbs.

2. Make sure leads are not worn, frayed, cut or show bare wire at any point. Do not join on extra wire to make leads longer. Fit a completely new longer lead (or use a proper extension lead).

3. Do not overload sockets. Double sockets and multisocket blocks are better and safer.

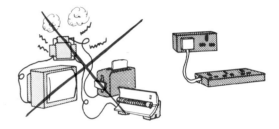

4. Do not run extension leads into the bathroom or use plug-in electrical equipment in the bathroom (except electric shavers in shaver sockets).

5. Do not trust sockets. Switches can go wrong and 'off' may still be 'on'. Pull out the plug before changing fuses or filling an electric kettle.

6. Do not poke anything into sockets or electrical equipment.

7 Reasons for the safety rules are given below. Work out which reason goes with which rule.

(a) The weight of two or three plugs on an adaptor can pull it partly out of a socket, exposing live metal prongs and make a bad connection. Too many high-power appliances in one socket will cause it to overheat or blow a fuse in the consumer unit.

(b) The effect of an electric shock on the body can vary greatly:
The skin is often wet in the bathroom and there should be no sockets or wall-mounted light switches in that room. An electric shock there could be fatal.

(c) Heating elements in toasters and electric fires are not insulated and are at a high voltage. Metal wires and knives used to poke the element will connect you to the mains. Ring main sockets are protected when not in use but small children should not be allowed to touch mains sockets at all.

(d) The lighting circuit is only designed for lamps. Other appliances, especially those that have heaters, may draw too much current through the cables. This will blow a fuse in the consumer unit, or over-heat the cable. The appliance will not be earthed.

(e) Constant use may cause bare wire to show through the insulation of a length of flex. The bare wire may be live and could shock you or start a fire. Exposed live and neutral wires may touch (blowing a fuse) or break (cutting off the electricity). Keep leads short and safe.

(f) You must be especially careful when handling electrical appliances with wet hands (something often done in the kitchen). The low resistance of the body when wet makes electric shocks particularly dangerous.

Body condition	Body resistance	Typical current	Effect of a shock
With a dry skin and thick insulation under foot	High 240 kΩ	small (1 mA)	The shock will give you a nasty jolt.
With sweaty skin and with part of the body earthed	Low 8 kΩ	large (30 mA)	The heart will fail and you could die.

Electric energy in circuits check list

After studying the chapters on Cells, Resistance, Power and Electricity in the home, you should know that:

- cells convert chemical energy into electrical energy
- that cells can be connected + to − to give more voltage
- that some cells can be recharged with chemical energy
- that the e.m.f. of a cell is the energy it gives to a coulomb of passing charge
- that a current only flows from a cell if there is a completely closed conducting path between its terminals
- that the same current flows through components connected in series
- that the current is shared by two components connected in parallel
- that resistance is the property of a conductor that limits the current that flows through it
- that the resistance of a wire increases with its length and decreases with increasing thickness
- that a potentiometer can be used as a variable resistor
- that a diode has a high resistance in one direction and a low resistance in the other.
- that an LED is a diode that emits coloured light
- that heat is produced when a current passes through a resistor
- that a fuse is a weak link in a circuit that melts and breaks the circuit if the current rises above its proper value
- that a 'short' is a low resistance path between the terminals of a voltage supply
- that an ammeter has a low resistance and must be connected in series in a circuit + to +
- that a voltmeter has a high resistance and must be connected across a component to measure its potential difference
- that the resistance of a component is measured by voltage drop/current (in ohms)
- that Ohm's law states that the resistance of a conductor is constant provided the temperature remains constant
- that the resistance of a thermistor goes down as it gets hot
- that the resistance of an LDR goes down as the light gets brighter
- that the resistance of an iron wire goes up a little as it is heated
- that the values of resistors add when in series
- and have a value of $R_1 R_2/(R_1 + R_2)$ when in parallel
- that a current divides at a junction and most goes the way of least resistance
- that two resistors in series across a voltage supply divide its voltage
- and the greater voltage difference is across the larger resistor
- that the electric power of a conductor is the amount of electrical energy it converts each second to other forms
- that electric power $P = VI$ and $P = V^2/I$ and $P = I^2 R$
- that a kilowatt hour is the energy used by a 1 kilowatt device in 1 hour
- that in a mains plug the brown wire is live, the blue wire neutral and the green/yellow wire the earth

Revision quiz

Use these questions to help you revise the work on Circuit electricity.

- What type of energy does a fully charged cell have? ... stored chemical energy.
- What is the e.m.f. of a cell or electric generator? ... the energy given to each coulomb of charge that passes through it.
- How is a rechargeable cell recharged? ... by driving current through it in the opposite way to its discharge current.
- What is meant by resistance? ... a conductor's ability to limit the current through it.
- Why must the same current flow out of a resistor that flows in? ... because if less flowed out there would be a build up of electric charge.
- How does the resistance of a wire depend on its dimensions? ... its resistance goes down with thickness and increases with length.
- Why must an ammeter have a low resistance? ... because the measured current passes through the ammeter and any resistance reduces the size of that current.
- Why must a voltmeter have a very high resistance? ... a voltmeter must draw very little current from the circuit so that it does not reduce the voltage it is measuring.
- What is p.d.? ... potential difference (or voltage difference) is the difference in electric potential energy at the ends of a conductor.
- How is the resistance of a conductor measured? ... voltage difference/current = resistance.
- What is Ohm's law? ... 'the current through a conductor is proportional to its applied voltage provided its temperature remains 'constant', or
  ... 'the resistance of a conductor is constant provided its temperature doesn't change'.
- How does the resistance of iron wire vary with temperature? ... it rises a little with temperature.
- And a thermistor? ... its resistance drops rapidly as the temperature rises.
- What is the special property of an LDR? ... its resistance drops rapidly as the light gets brighter.
- What is the special property of a diode? ... it has a high resistance in one direction and a low resistance in the other.
- What is an LED? ... a diode that emits coloured light when it conducts.
- What is the resistance of R_1 and R_2 in series? ... $R=R_1+R_2$.
- What is the resistance of R_1 and R_2 in parallel? ... $R=R_1 R_2/(R_1+R_2)$.
- What is a voltage divider? ... two resistors (R_1 and R_2) in series across a voltage supply (V). The voltage divides into $$\frac{R_1 \times V}{(R_1+R_2)} \text{ and } \frac{R_2 \times V}{(R_1+R_2)}.$$
- What is the electric power of a component? ... the energy converted/second from electrical to other forms. Measured in joules/second or watts.
- What are the equations for the power of a resistor R? ... $P=VI$ $P=V^2/R$ $P=I^2R$.
- What instruments are needed to measure power? ... a voltmeter and ammeter or a joulemeter and clock.
- What is a kilowatt hour? ... the energy used by a 1 kilowatt device in one hour.
- What are the colours of the wires in mains flex? ... blue (neutral), brown (live), green/yellow (earth).

Examination questions

1. The diagram illustrates an arrangement for measuring the energy input and the energy ouptut of an electric motor. It is found that the 6 kg mass, initially at rest, can be raised 3 m in 5 s and that the readings of the ammeter and voltmeter are 2 A and 24 V respectively. (The gravitational field strength is 10 N/kg.)
   (a) How many coulombs of electricity have passed through the motor?
   (b) What is the energy input from the supply?
   (c) How much energy was used in lifting the 6 kg mass?
   (d) It is found that the gain in energy of the mass is never equal to the energy put in by the electrical supply. Account for this difference in energy.
   (MEG)

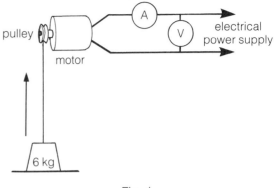

Fig. 1

2. This question is about the voltages across devices connected in series.
   (a) Fig. 2 shows a 500 ohm resistor and a 1000 ohm resistor connected in series across a constant 3 V d.c. supply.
      (i) What is the total resistance across the supply?
      (ii) What is the current through the circuit?
   (b) When a voltmeter is connected across LN, as shown in Fig. 3, it shows a voltage of 3 V.
      The voltmeter is then connected across MN as shown in Fig. 4. What voltage will it now show?
   (c) The 1000 ohm resistor is taken out of the circuit and a well-lit light-dependent resistor (LDR) is put in its place, as shown in Fig. 5.
      (i) The voltmeter now shows a voltage of 1.5 V. What is the resistance of the LDR?
      (ii) What would happen to the voltmeter reading if no light were allowed to fall on the LDR (by wrapping it in black cloth for example)?
      (iii) Explain why this would happen.
   (SEG)

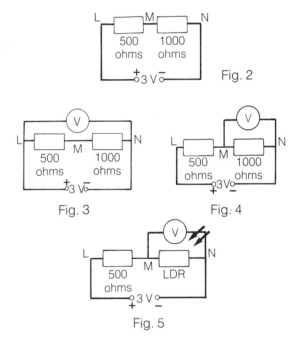

3. (a) (i) Why is a fuse fitted in a modern 3-pin plug?
      (ii) Explain why it is safer to use a 3 A fuse in the plug for a table-lamp rather than the 13 A fuse often sold with the plug.
      (iii) One of the pins is connected to the 'earth'. For an appliance with a three-core cable, describe the purpose of the 'earth' and explain how it improves safety.
   (b) An electric fire is rated 250 V 1 kW. The fire is operating normally.
      (i) How much electrical energy is converted in 1 second?
      (ii) Calculate the current in the element.
      (iii) Calculate the resistance of the element.

(c) An aircraft flies just below a negatively-charged thundercloud. Movement of free electrons causes electrostatic charges to be induced in the aircraft.
  (i) Copy and draw on the diagram the positions and signs of the induced charges on the aircraft.
  (ii) Explain in terms of electron movement, the distribution of charges you have drawn.
  (iii) What will happen to the induced charges when the aircraft flies away from the cloud?

(*LEAG*)

Fig. 6

4 A six-volt battery is used with a potential divider to provide a variable voltage across a torch bulb rated at 3 V 0.3 A.
  (a) Sketch a circuit diagram showing
    (i) the arrangement of the potential divider and bulb.
    (ii) how the voltage across the bulb could be measured.
    (iii) how the current through the bulb could be measured.
  (b) Sketch a graph showing how the current through the bulb varies with the voltage across it.
  (c) Describe and explain how the resistance of the bulb filament changes with increasing current.

(*NEA*)

5 Fig. 7 shows a diagram of an electrical circuit.
  (a) Name the device P.
  (b) What effect does adjusting P have on the lamp?
  (c) Why does P have this effect?
  (d) Copy and add to Fig. 7.
    (i) an ammeter to measure the current (use the symbol Ⓐ).
    (ii) a voltmeter to measure the potential difference across the lamp (use the symbol Ⓥ).
  (e) Draw a circuit diagram of your completed circuit, including the ammeter and voltmeter.

(*NEA*)

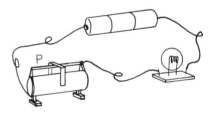

Fig. 7

6 Fig. 8 shows a diagram of a fused three-pin plug. The plug is connected to an electric fire.
  (a) What is the colour of the insulation of the wire connected to
    (i) pin X, (ii) pin Y?
  (b) Why is pin X longer than the other two pins?
  (c) What is the purpose of R?
  (d) Wire X is connected to the frame of the fire. Why does this protect the user from electric shock?

(*NEA*)

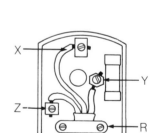

Fig. 8

7 (a) (i) Describe with the aid of a circuit diagram an experiment to measure the resistance of a provided length of wire.
  (ii) What additional measurements would have to be made in order that the resistivity of the material of the wire could be determined.

(b) An experiment is carried out to measure the potential difference across a filament bulb and the corresponding currents which pass through it. The following results were obtained:

Voltage (V)	0	1	2	3	4
Current (A)	0	0.21	0.34	0.43	0.50

  (i) Use these results to draw a graph of voltage against current.
  (ii) From the graph determine the resistance of the filament when 0.5 volt is applied and when 4.5 volts is applied.
  (iii) Suggest a reason why the values are different.
(c) Ten identical bulbs, each with an operating resistance of 10 Ω are connected in parallel to a 10-volt battery.
  (i) What is the total current drawn from the supply by the bulbs?
  (ii) How much electrical power does *each* bulb consume?

(*NISEC*)

You are provided with a cubic block of aluminium, a heater of unstated power and a thermometer.
The block is surrounded by a polystyrene insulation and contains two holes which hold the heater and the thermometer.
(a) Copy and draw on the diagram below an electrical circuit which allows you to determine the electrical power supplied to the heater in order to raise the temperature of the block. You are provided with a suitable d.c. voltage supply, a voltmeter, switch, ammeter and connecting leads.
(b) How would you calculate the total energy supplied in order to raise the temperature of the block by about 20 °C?
(c) What additional measurement would be needed to find a value for the specific heat capacity of aluminium?
(d) Would you *expect* the value which you get for the specific heat capacity to be higher or lower than the accepted value? Explain your answer briefly.
(e) The insulation and heater are removed when the block reaches its maximum temperature. The room temperature is 20 °C and the block has a temperature of 40 °C. The temperature of the block is recorded at half-minute intervals and the results are plotted on the axes shown below. Sketch the shape of the graph which you would expect.
(f) Outline how you could use the graph to compare the rate of loss of heat by the cube initially and after 3 minutes.
(g) A second cubic block of aluminium has sides which are twice as long as the original block.
  Copy and complete the table below.

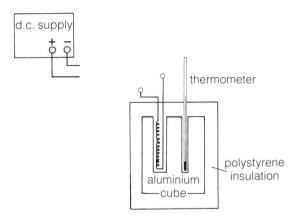

Fig. 9

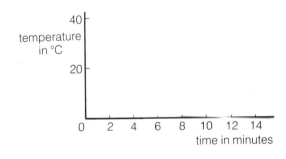

Fig. 10

	Side	Total surface area	Volume	Mass
Block 1	l	A	V	M
Block 2	2l			

(h) If both blocks start at the same temperature (40 °C), sketch the general shape which you would expect for graphs of temperature against time for *both* blocks.
  Clearly label the graphs 'block 1' and 'block 2'.
(i) If your graphs for the two blocks are not the same, suggest a reason for the difference.

(*NISEC*)

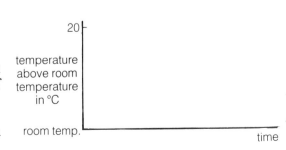

Fig. 11

25 Magnetism

EXPERIMENT. Finding the poles of a magnet

Roll a magnet in iron filings. You will find that the filings stick mostly round the ends of the magnet. These places, where the magnetism is strongest, are called the **poles** of the magnet. How many poles has your magnet got, and where are they?

The poles of a magnet

EXPERIMENT. Are both magnetic poles the same?

The two poles of a bar magnet are near its ends and they both attract iron filings but are they different in any way? Hang a magnet on a thread so that it can swing. You will see that one end turns until it points (roughly) to the North pole of the Earth. If you point this end to the South, it will turn round again until it points North. This pole is called the North pole of the magnet (N). The other end that points South is called the South pole (S). Write N and S on the poles of your magnet.

Finding the N pole of a bar magnet

Forces between magnetic poles

EXPERIMENT

Place two magnets on some polystyrene beads (so they can slide easily) with N poles together. Let go and the magnets fly apart. Each magnet pushes the other away (repels it) even when they do not touch. S poles repel each other too but S and N poles attract.

Magnets repelling each other

1 Copy this table and write 'repel' or 'attract' to say what happens when the poles meet.

	North	South
North		
South		

2 These two magnets repel when placed like (a). Will they attract or repel when placed like (b)?

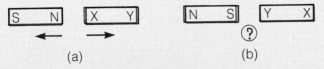

(a)   (b)

3 The needle of a compass is a magnet that can spin on a point.
  (a) Should the arrow of the needle be a N or S pole?
  (b) Sometimes the needle gets magnetized the wrong way round. How would you check which end of the compass needle was the N pole?

4 The two compass needles set as shown when placed near the magnet. What type of poles are X and Y?

5 A student placed a compass under wire in an electric circuit. When the current was switched on, the compass needle no longer pointed North.
  (a) What does this experiment suggest about electric currents?
  (b) What could you do to make sure that light from the lamp is **not** causing the movement of the compass needle?

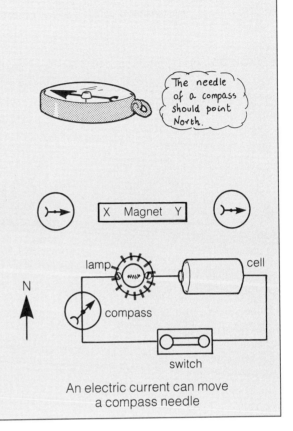

An electric current can move a compass needle

The Earth is magnetized

The N pole of a magnet that can swing freely will point to the Arctic. Since N poles are attracted to S poles, the Arctic must be like a S pole and the Antarctic a N pole. The Earth acts like a magnetized ball with its magnetic poles near the geographical poles.

EXPERIMENT

Make a model of the 'Earth magnet' by burying a bar magnet in a ball of plasticine. Roll the ball in iron filings to show up the magnetic poles and use a small compass to show that compass needles point to the Arctic wherever they are.

EXPERIMENT. Making a steel magnet

Start with a piece of steel that is not magnetized (test it by dipping it in iron filings). Slide the N pole of a strong magnet along the steel many times in the same direction. The steel should become a magnet. Check by dipping it again in iron filings. Does end X become a N or S pole? Use a compass needle to find out*. (A S pole will attract the point of the compass needle and repel its tail.)

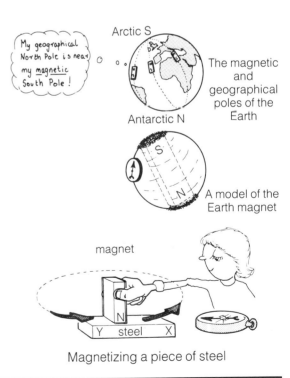

Magnetizing a piece of steel

* X becomes a S pole.

EXPERIMENT. Can you make an iron magnet?

Use the method on p. 247 to try and magnetize iron (a paper clip behaves something like iron) and a pen nib or screwdriver blade (steel). Use iron filings to find which is made into a 'full-time' (permanent) magnet. When you tap the iron the filings fall off, but when you tap the steel nib some filings stay on.

6 Copy this table and insert yes or no into the spaces.

Iron	Steel	
		Is it attracted to a magnet?
		Can it be made into a permanent magnet?

This experiment shows again the magnetic difference between iron and steel. When on or near a magnet, iron becomes a strong magnet. But when the magnet is removed the iron loses its magnetism. A steel nib or sewing machine on the other hand holds on to its magnetism after the magnet has been removed.

Alloys containing nickel and cobalt can also be magnetized strongly and like steel hold onto their magnetism.

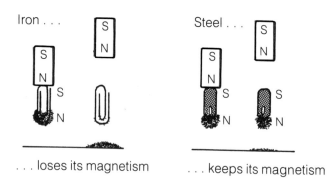

... loses its magnetism   ... keeps its magnetism

7 The diagram shows paper clips and nibs hanging from a magnet. The two chains are removed by pulling the first clip away from the magnet. Which chain will fall apart and which will stick together? Explain your answer.

8 A magnet attracts a metal bar X whichever pole is used to do it. The same magnet attracts and repels a metal bar Y depending on which pole is used. Which bar is a magnet, and which is a piece of iron?

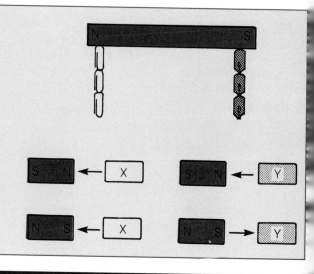

9 Why is repulsion the only sure test that a metal bar is a magnet?

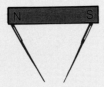

10 Explain why needles; hung from the ends of a bar magnet, lean towards each other as shown.

11 Magnets keep their magnetism longer if they are stored in pairs with iron 'keepers' across their ends. Explain why magnets arranged like this are not very good at picking up iron filings.

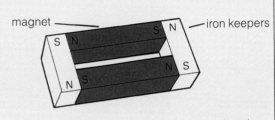

Storing magnets with iron keepers so that they do not lose their magnetism

Magnetic fields

The influence of a magnet can be detected in the space around it. A force acts on pieces of iron and steel in that space. A magnet is surrounded by an invisible **magnetic field**. Iron filings can be used to show up the shape of the magnetic field.

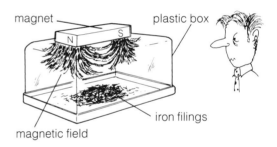

EXPERIMENT. Looking at magnetic field patterns

Sprinkle a fine even coat of iron filings on a sheet of paper that is covering a bar magnet. Tap the paper gently and the filings will move into a pattern that shows the shape of the magnetic field. Make a drawing of the magnetic field.

The iron filings can be fixed to the paper by spraying them gently with hair spray.

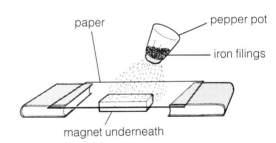

Using iron filings to show up a magnetic field

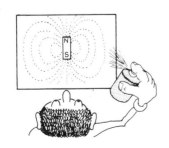

Magnetic field lines

EXPERIMENT. Exploring magnetic fields with a small compass

A sensitive way of revealing the pattern of a magnetic field is to use a small 'plotting' compass.

Lay a piece of paper in the magnetic field, choose a starting point and mark it with a pencil dot. Put the tail (S pole) of a plotting compass over the dot and draw a second dot at its point (N pole). Move the compass along so that its tail covers this dot and continue.

Dot the path of the N pole in this way as it leads you through the magnetic field. Then join the dots to get the line followed by the N pole in the field. This line is called a **magnetic field line**.

Follow other lines by starting at different points in the field until you have a pattern that shows the shape of the magnetic field.

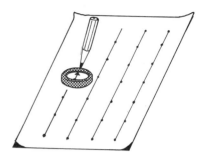

Note that magnetic field lines:
(a) go from the N pole of a magnet to its S pole and have an arrow to show this direction;
(b) show the direction of the force on a N pole;
(c) are close together where the field is strong;
(d) never cross.

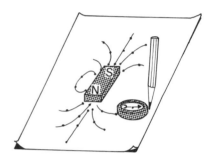

> 12 Give a reason why magnetic field lines cannot cross. What would happen to a compass placed at that point? Use compass plotting to show the shape of other magnetic fields you have designed.

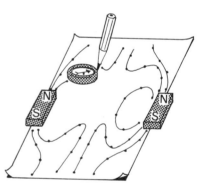

Pieces of iron change the shape of magnetic fields

Whenever a piece of iron is laid in a magnetic field, it disturbs the field and alters the shape of the field lines.

EXPERIMENT

Place a piece of soft iron close to a strong magnet. Use compass plotting to trace the field lines around the magnet and iron. Then trace the field lines without the iron. Use the two patterns to show how the iron disturbs the magnetic field.

The new direction of the field lines may show that the iron has gained N and S poles and has become a magnet itself. Magnetism, forced onto the iron like this, is said to be induced.

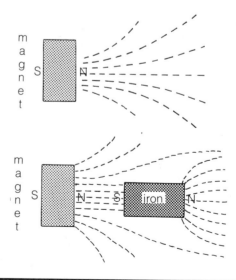

> 13 What would happen to the iron if it were free to move?

Using iron to shape magnetic fields

The pictures below show magnetic fields before and after pieces of iron have been added to strengthen or alter their shape.

14 The magnetic fields have been shaped for the following jobs:
 (a) a loudspeaker magnet,
 (b) an electromagnet,
 (c) a radial field for an electric motor or galvanometer,
 (d) an intense field for research.

Which picture shows the magnet used for each of these purposes?

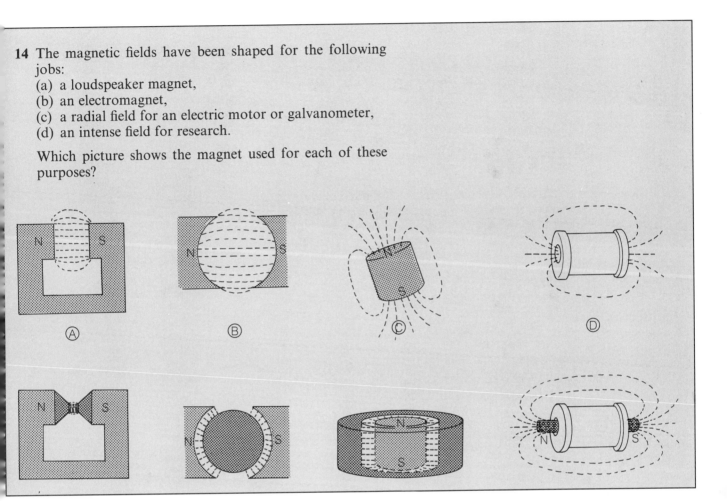

26 Electromagnetism

An electric current and its magnetic field

EXPERIMENT. Magnetism from electricity

Use a compass to see if there is a magnetic field near to an electric current.

Take a long length of insulated wire and wind it carefully around a compass. Connect the ends of the wire into a circuit with a lamp, switch and voltage box. Turn the compass so that the needle lines up with the coil. Switch on and look for a movement of the needle.

Do you agree that
(a) an electric current in a coil does make a magnetic field,
(b) the field gets stronger if the current gets larger?

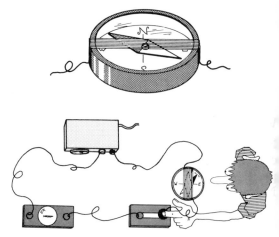

> **1** Describe how you could make a simple ammeter from a compass and a length of insulated wire.

The shape of the magnetic field of a current

EXPERIMENT. The magnetic field of a current in a long straight wire

Pass the electric current along straight wires that go through the middle of a board. Sprinkle iron filings on the board and tap gently. The filings will show that the shape of the field is circular.

Instead of filings use a ring of plotting compasses. Switch the current on, and the needles will show the direction of the field around the wire.

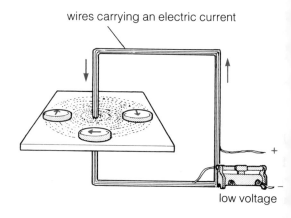

wires carrying an electric current

low voltage

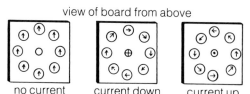

view of board from above

no current   current down   current up

Looking at the magnetic field of an electric current

252

EXPERIMENT. The magnetic field of a current in a solenoid

A **solenoid** is a length of wire wound into a long coil.

Make a card that fits around a solenoid and sprinkle iron filings on the card. Pass a current through the solenoid and look at the pattern that the filings make.

Outside the solenoid the magnetic field should look familiar. It is the same shape as the field of a bar magnet. Inside, the field is concentrated and uniform in strength. The magnetic field makes the ends of the solenoid behave like N and S poles.

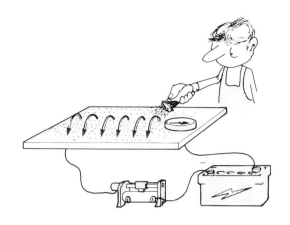

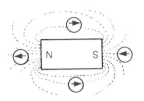 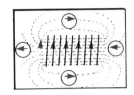

Which end is North?

Here is a handy rule for finding the N ends of solenoids when they carry current. Grab the solenoid in your right hand, with the fingers pointing the same way as the current. The thumb then points to the N end of the solenoid.

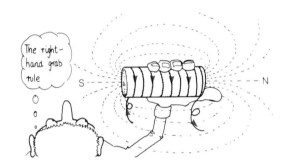

2 A student built this apparatus to see if solenoids really do behave like magnets.
   (a) The S pole of the magnet attracts end A of the coil and pulls the floating coil across the water. Name two changes that would make the coil move away from the magnet.
   (b) If the magnet is fixed to the coil would it drive the coil along?

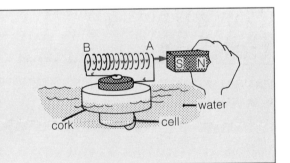

Uses of solenoids

Magnetizing

EXPERIMENT. Making a magnet

The strong magnetic field inside a solenoid carrying a current can be used to make steel into a magnet.

Use a hollow solenoid made from many turns of wire. Connect it into a circuit with a switch and voltage box. Place the steel inside the solenoid and switch on the current. The field passes through the steel and magnetizes it.

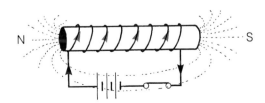

Using a solenoid and current to magnetize steel

253

From your experiment do you agree that:
(a) the steel keeps its magnetism after the current is switched off,
(b) the N pole is at the end predicted by the right-hand-grab rule,
(c) an alternating current in the solenoid can magnetize the steel,
(d) an iron core becomes magnetized, but loses its magnetism when the current is switched off.

3 Which end of the bar becomes a N pole?

4 An alternating current (current that keeps changing direction rapidly) can be used to magnetize steel, but you can never tell which end will be a N pole when you switch off. Can you explain this?

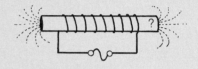

5 When the switch in this apparatus is pressed the solenoid magnetizes and pulls the magnet down. If you press in rhythm with the bouncing magnet, large amplitude bounces build up.
(a) Should the top end of the solenoid become a N or S pole for this to work?
(b) Is the battery in the diagram connected correctly to do this?
(c) Give another example of small regular forces being used to build up large amplitude vibrations (resonance).

EXPERIMENT. Making an electromagnet – a magnet that can be switched on and off

Use about 2 m of plastic covered wire to wind a solenoid around a piece of iron. Strip the plastic from the ends of the wire and connect them into a circuit with a switch and a voltage box. If the iron picks up filings when the current is on and drops them when it is off, you have succeeded in making an electromagnet.

Find out if the electromagnet
(a) gets stronger when you use more turns of wire,
(b) gets stronger when the current is larger (turn up the voltage a little),
(c) works with alternating current.

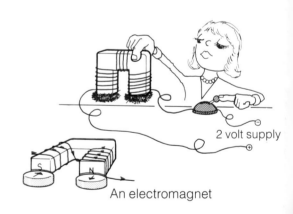

An electromagnet

6 Why is iron (and not steel) used for electromagnets?

7 Two electromagnets were made, one with an iron core and one with a steel core. What will happen when the current is switched on and off?
Copy the table and put 'magnetized' or 'unmagnetized' in the boxes.

	Iron	Steel
Current on		
Current off		

8 Why must the wire for electromagnets be insulated?

9 An electromagnet holds up a steel ball.
   (a) Give two ways of making the electromagnet stronger.
   (b) The ball does not fall off when the magnet is switched off. Explain why not.

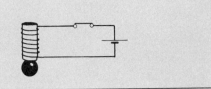

Demagnetizing

EXPERIMENT. Demagnetizing a magnet

An excellent way of getting rid of magnetism is to use a solenoid carrying an alternating current.

Put the magnet inside the solenoid, switch on the current and then pull the magnet out until it is well clear. You will feel vibrations as the tiny magnets in the metal are jumbled up by the magnetic field of the alternating current. Use iron filings to test that the magnetism has gone.

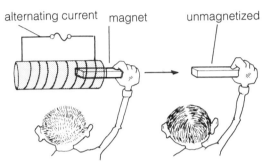

Using a solenoid and alternating current to demagnetize a magnet

10 Describe how you can use a solenoid to
   (a) magnetize,
   (b) demagnetize a piece of steel.

Uses of electromagnets

Reed switches

A reed switch has two iron reeds sealed inside a glass tube. The switch is closed by bringing up a magnet. This magnetizes the reeds so that they attract each other and make contact. The current can then flow between them. There are two types: (i) normally open – a nearby magnet closes the contacts; (ii) normally closed – a nearby magnet opens the contacts.

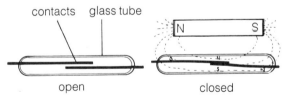

Reed switches

11 Copy and write 'open' or 'closed' in the spaces.

	Magnet near	Magnet nowhere near
Normally open reed switch		
Normally closed reed switch		

255

EXPERIMENT

Build this circuit and light the lamp by bringing up a magnet. Can you explain why it is best to hold the magnet parallel to the reed switch?

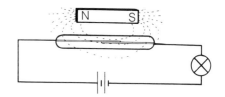

12 How to make an automatic door bell. Fix a magnet on the door and a reed switch to the door frame. Connect the reed switch to a bell and battery. When the door opens, the magnet moves away from the reed switch allowing it to close and ring the bell.
   (a) Is a normally open or normally closed reed switch needed for this?
   (b) Why is this arrangement not a very effective burglar alarm?

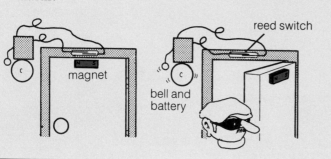

Reed relay

A current in a solenoid can be used instead of a magnet to switch on a reed switch. Closing the switch S produces a magnetic field in the solenoid. This closes the reed switch and switches on the lamp. Note there are two quite separate circuits. This arrangement is called a **relay**.

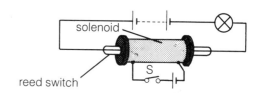

Electromagnetic relay

Closing switch S in this relay will magnetize the iron core. This core will then attract A, closing the lamp circuit and lighting the lamp. A small current in the solenoid can switch on a large current in the lamp circuit.

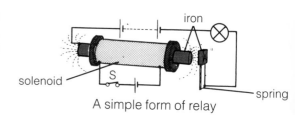

A simple form of relay

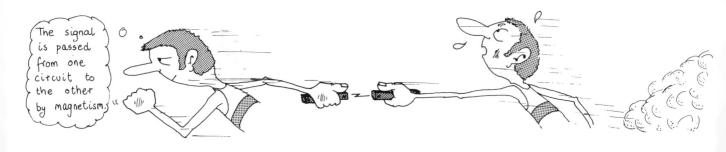

The use of a relay in a car

The starter motor in a car uses a very large current (up to 100 A). A relay is used so that this large current can be switched on by a small current. The small current flows through the starter switch and into a solenoid. The solenoid magnetizes and pulls in a piece of iron, closing heavy duty contacts between the battery and starter motor.

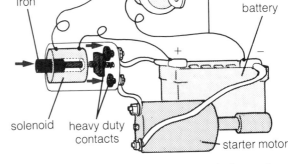

Using a relay to switch on a car starter motor

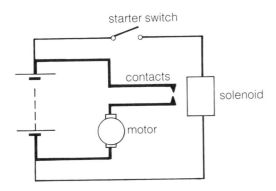

13 Explain why the leads to the starter motor are very thick and are kept as short as possible.

Relays in electronic control

Relays are widely used to make electronic circuits (that use small currents) switch on things such as motors and heaters that use large currents.

Here are some of the types of relay that are used. They all have a solenoid (or coil) that is connected to the 'small current' circuit, and contacts that are used to switch large currents in a separate circuit.

EXPERIMENT. Electronic control using a relay

This circuit allows a light beam to switch on a motor. A light-dependent resistor (p. 221) is used to sense the light and a relay is used to switch on the motor.

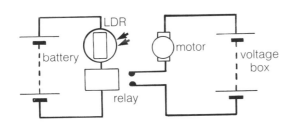

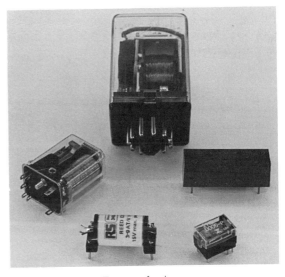

Types of relay

257

Solder four wires to the solenoid pins and switch pins of a relay. Connect the solenoid wires to an LDR and a 9 V battery, and the switch wires to the motor and a voltage box. A bright light on the LDR should now start the motor.

Do tests on your circuit to show the following:
(a) the motor will not start if it is put in a circuit with just the battery and the LDR.
(b) the current in the LDR circuit is much smaller than the current in the motor circuit.

This circuit does the same job but only uses one voltage supply. Build it and check that it works as well as before.

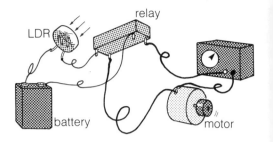

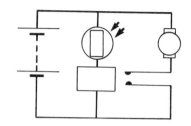

> **14** Copy the circuit diagram for this arrangement and mark clearly the 'small current circuit' and the 'large current circuit'.
>
> **15** Explain why shining light on the LDR starts the motor. Say what happens
> (a) to the resistance of the LDR,
> (b) how this changes the current through the solenoid,
> (c) and what this does inside the relay.

Buzzers and bells

An electric buzzer can be made from an electromagnet and a switch that opens and closes automatically. Look at the diagram of the buzzer and follow the current round its circuit. The current has to pass between a screw (S) and an iron piece (A). The diagrams below show how the current switches itself on and off.

Start here
When A touches the screw a current flows, and the electromagnet switches on. A is attracted to the magnet and moves away from S. A gap appears between A and S. The current switches off, the electromagnet loses its magnetism and the spring returns A to the screw S. (Go back to 'start'.)

This to and fro motion happens very quickly (about 10 times a second) and makes a buzzing noise. In an electric bell a hammer attached to A hits a gong and makes it ring.

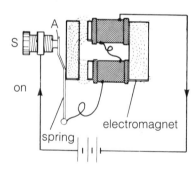

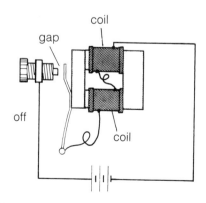

16 Describe in your own words how the buzzer works.

17 Why must piece A be made of iron and not steel? What would happen if it was made of steel?

18 Would the buzzer work with the wires connected as shown in this diagram? What happens when you switch on?

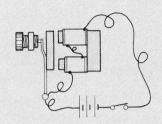

19 The diagram shows a bell with an electromagnet that can make a hammer hit two metal plates. When struck, the plates go 'ding' and 'dong'. The switch is closed and then opened again. Would the bell go 'ding-dong' or 'dong-ding'?

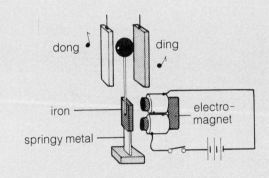

20 Copy and complete

Part of bell	Material used	Reason for using that material
The core of the electromagnet		
The wire of the coil		
The spring		

21 This diagram is of an electric bell.
   (a) Name the parts A and B.
   (b) Name the materials used to make parts C, D and E.
   (c) Explain why the bell vibrates when the switch is closed.

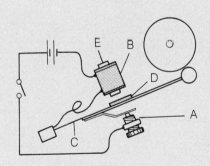

259

27 The electric motor force

Electricity and magnetism together can produce a force. Electric motors make use of this force to turn loads. They convert electric and magnetic energy into work.

EXPERIMENT. Investigating the motor force

You can see the motor force in action if you pass a current through a strip of aluminium foil that lies between the poles of magnet.

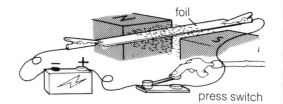

Connect the strip into a circuit with a resistance, press switch and voltage box. Hold the strip between the poles of the magnet and switch on the current. The motor force acts on the current and makes the strip jump out of the field.

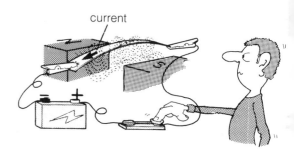

Do experiments to show that
(a) the force acts the other way if the current is reversed or if the magnet poles are reversed;
(b) the force gets weaker if the current is made smaller;
(c) there is no force if the strip lies along the magnetic field lines;
(d) alternating current through the strip makes it move rapidly up and down.

1 The strip is folded into a loop and placed between the magnet poles. Which way will the forces act on each side of the loop? How will these forces move the loop? (Try this with your apparatus to see if you are right.)

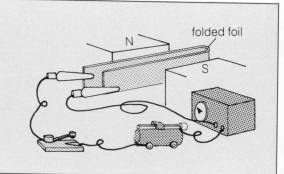

The left-hand rule

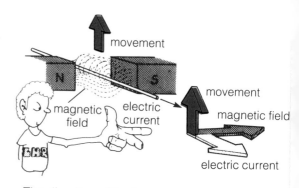

The direction of the force on an electric current in a magnetic field

The movement, the electric current and the magnetic field are all in different directions. You can use your left hand to work out which way the foil (or a wire carrying a current) will move.

Point the First finger in the direction of the magnetic Field (N to S),
    the seCond finger in the direction of the electric Current (+ to −),
        the thuMb shows the direction of Movement of the conductor (or the force on it).

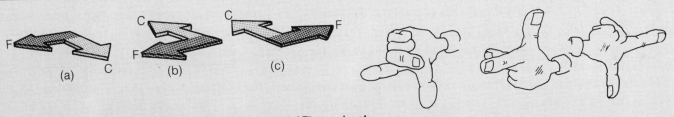

2 If the magnetic field (F) and electric current (C) are in the directions shown by the arrows, use your left hand to work out whether the movement of the conductor is up or down.

3 This picture shows the magnetic field of a current flowing through the magnetic field of a magnet. Explain
(a) why the magnetic field is stronger under the wire than above.
(b) which way you would expect the wire to move.

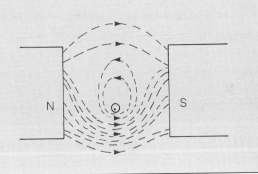

The parts of an electric motor

Look closely at a d.c. motor and see if you can find the following parts:

1. The rotor, made of turns of insulated copper wire, that can rotate on an axle.
2. The commutator, on the end of the rotor, made of strips of copper.
3. The brushes, that press onto the commutator and are connected to the current terminals.
4. The magnet that surrounds the rotor.

4 This table gives the materials that can be used for the parts of a simple motor. Give a reason for using each material.

Part	Materials used	Reason
Rotor coil	enamelled copper wire	
Brushes	carbon	
Magnet	steel	

A d.c. electric motor

How the motor works

If an electric current is passed through two wires that lie between the poles of a magnet, an upward force will act on one and a downward force on the other. If the wires make a loop that can spin on an axis, the two forces will turn the loop. It will not go far though because the wires from the battery will get twisted and after half a turn the forces will flick the loop back.

The loop can be made to spin by fixing a half circle of copper strip to each end of the loop. (These copper strips are called a **commutator**.) Current is passed in and out of the loop by 'brushes' that press onto the commutator strips. The brushes do not go round and so the wires do not get twisted. This arrangement also makes sure that the current always passes the magnets in the same directions so that the rotation continues. This is why an electric motor spins when it is fed with current.

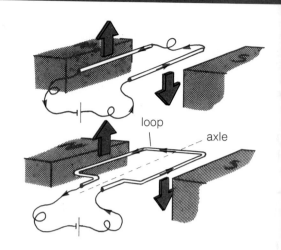

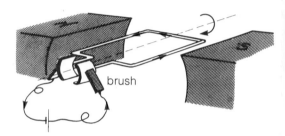

A simple electric motor

Experiments with an electric motor (see also p. 87)

Connect an electric motor, ammeter and switch to a voltage box. (You will find the voltage to use written on the motor.) Switch on and find answers to this questionnaire.

What voltage does the motor use?	
How much current does it use when running freely?	
What power does it use when running freely?	
How much current does it use when driving a load? (Press a rubber on its axle or make it lift a weight.)	
What power does it use when driving a load?	
How many revolutions does it make in one minute when running freely (use a rev. counter)?	
How many revolutions does it make in one minute when driving a load?	

Examining a low-voltage electric motor

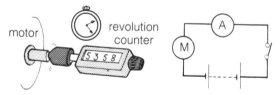

Finding the number of revolutions/minute

EXPERIMENT A speed control for the motor

Include a potentiometer in the circuit with the motor and the ammeter. Moving the potentiometer slider will change the current going to the motor and its speed. Use a revolution counter to measure the speed of the motor (in revs/second) for different current values.

Note there are different types of electric motor that have other ways of controlling their speed.

Controlling the speed of a low-voltage motor

28 Electricity generators

How is the electricity made that we use at home? It is not produced by batteries, but by dynamos and alternators.

Dynamos – direct current generators

EXPERIMENT. Looking at the current from a dynamo

Connect a dynamo to an ammeter that has zero in the centre of its scale. Turn the dynamo and you will see the pointer move out and back on one side of the zero. The current flows one way round the circuit and rises and falls in value. Compare this with the current from a cell, which flows one way and has a steady value. Both are called direct currents because they flow in only one direction.

A dynamo has the same parts as a d.c. electric motor. It has a coil of insulated wire that can rotate on an axle between the poles of a magnet. The ends of the coil are connected to commutator segments. Brushes press on the segments to take the current to the outside circuit. Current is generated when the wires of the coil move past the poles of the magnet.

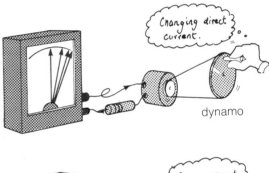

Steady and changing direct current

A d.c. dynamo

1 If you turn a d.c. motor by hand it generates current. If you supply current to the motor it produces movement. Which of these arrows shows the energy changes in (a) an electric motor, (b) a dynamo? Why can't the current generated be fed back to keep the motor running?

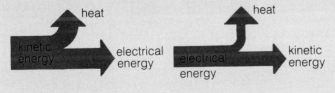

EXPERIMENT. A closer look at how the current is generated

The simplest way of making a current is to move a length of wire between the poles of a strong magnet.

Connect the ends of the wire to a sensitive ammeter to show up current flow. You can make the current larger by winding a length of wire into a large loop and moving one side of the loop through the magnetic field.

Use your apparatus to check the following observations.

The current flows the other way if you move the wire up instead of down.	
There is no current when there is no movement or no magnetic field.	
A larger current flows if you move the wire faster.	
A current flows if the wire is held still and the magnet is moved.	

(Tick the statements you find to be true.)

> **2** In this experiment where does the energy come from to make the current?
>
> **3** Which two of these changes will probably make the current larger:
> (a) using thinner wire,
> (b) using stronger magnets,
> (c) moving the wire faster,
> (d) using a more sensitive meter.

The right-hand rule

The direction of the current produced by the movement and magnetism can be found using the fingers of the right hand. The fingers are used in the same way as the left-hand rule (p. 260).

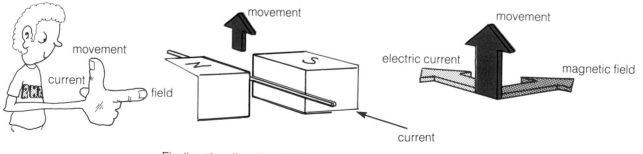

Finding the direction of the generated current

Alternators – alternating current generators

A simple alternator is made of the same parts as a dynamo except that two circular 'slip' rings are used instead of two half-rings.

It generates a current that first flows one way round the circuit, stops and then flows the other way round. This is called **alternating current (a.c.)**. It lights a lamp almost as well as direct current but makes the pointer of an ammeter swing from + to − to + each turn of the loop.

Connect an alternator to a centre-zero ammeter and look at this current for yourself.

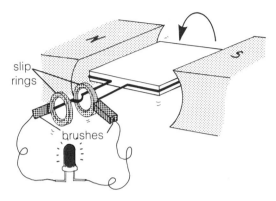

A simple alternator

4 Copy this table that describes the current produced by dynamo, alternator and cell. Draw these graphs alongside the description that it fits.

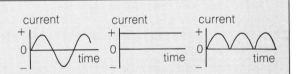

Description of the current	Graph of the current against time	What produces current like this?
The current gets larger then smaller until it dies away altogether. The current then continues to rise and fall in size, always flowing one way through the wires.		
The current stays the same size and flows the same way all the time.		
The current gets larger then smaller until it dies away altogether. It then begins to flow the opposite way, getting larger and smaller in the other direction.		

5 Copy and complete.

Source of current	Is the current steady?	Does it flow one way only?	Does it flow both ways?	Is the current changing in size all the time?	a.c. or d.c.?
Cell					
Dynamo					
Alternator					

6 An electricity generator is connected to an electromagnet. How would you use a compass needle to find whether alternating or direct current is being produced?

7 Suggest one way of making a dynamo generate more electric current.

8 What type of current is 'mains current', alternating or direct current?

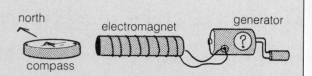

Practical generators of electricity

Alternator

The current produced by an alternator can be increased by winding a loop of many turns of wire onto an iron core and curving the poles of the magnet. Sometimes electromagnets are used instead of permanent magnets. Electromagnets can be stronger but need a supply of direct current to work them. Alternators are found in some car and lorry engines.

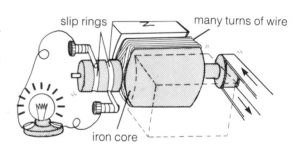

Cycle alternator

We have seen that current is generated if the magnet is moved instead of the wire (p. 264). Sometimes it is more convenient to move the magnet. Brushes and slip rings are not needed if the coil does not move. The cycle alternator shown here has a circular magnet that is made to spin inside an iron core. Alternating current is generated in a coil wound on this core.

Mains alternator

The alternators that produce 'mains' electricity also have a rotating magnet inside stationary coils. The rotating magnet is an electromagnet that is supplied with direct current from a dynamo on the same drive shaft. The electromagnet and dynamo are kept spinning at exactly 50 revolutions/second by a steam turbine (p. 143). Alternating current is generating in the stationary coils and can be led to our homes without brushes or slip rings.

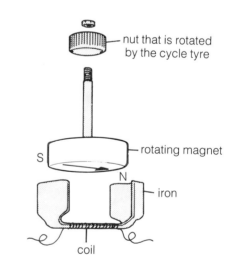

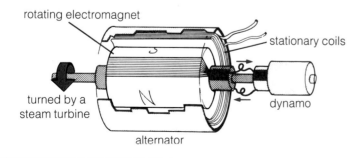

Generator speed and voltage

Experiment with a bike alternator

You will have found that an alternator produces more voltage when it is turned faster. This experiment looks for a connection between speed of rotation of a bike alternator and the voltage it produces.

Balance your bike upside down on the bench and disconnect the lamp from the alternator. Connect an a.c. voltmeter to the alternator and use the pedals to rotate it at a steady rate. Hold the output steady at 1 V for 1 minute and count the number of pedal turns in this time.

Increase the speed until the output is 2 V and again count the number of pedal turns in 1 minute. Carry on like this until you can go no faster.

To look for a connection between voltage and rotation speed, plot a graph of the output voltage against pedal turns per minute. A straight line through the origin shows that voltage and rotation speed are proportional.

What do your results show?

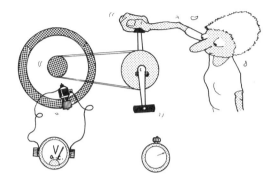

9 One turn of the pedals produces many more turns of the alternator magnet. How would you measure the number of turns of the alternator for 1 turn of the pedals?

10 These results were obtained from an experiment like this except that a frequency meter was used to measure the rotation rate of the alternator. Plot a suitable graph to find out if output voltage and rotation rate are proportional.

Output voltage (V)	2.0	3.3	4.4	6.3	7.4	8.8	10.0	13.0
Magnet rotation frequency (Hz)	32	55	75	100	120	147	163	210

EXPERIMENT. Using an oscilloscope to look at the alternator voltage

See if you can get a steady picture of the voltage from the bike alternator on an oscilloscope screen.

Connect the wires from the alternator to the terminals of the oscilloscope and then turn the pedals at a steady speed. Adjust the oscilloscope controls (p. 271) until the voltage wave is steady on the screen.

Find out what difference it makes to the shape of the voltage wave when you turn the alternator faster.

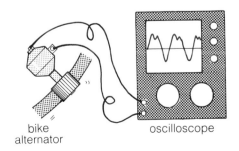
bike alternator   oscilloscope

11 These oscilloscope pictures show the voltage wave from fast-moving and slow-moving alternators. Describe and explain two differences between them. How can you tell whether the voltage is alternating or direct?

 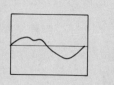

Changing a.c. to d.c.

Most electronic devices use direct current so it is useful to change the alternating current from alternators to direct current.

We have seen that a diode is a one-way device or valve for current. It can be used to change alternating current into direct current.

12 Here are some one-way devices. Copy the table and write what it is that can flow one way but not the other.

13 Draw these circuits and put ☼ on lamps that light.

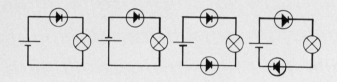

Turnstile	
Tyre valve	
Heart valve	
Diode	

EXPERIMENT. Slow-motion rectification

Connect an alternator to a diode, lamp and centre-reading ammeter. Turn the alternator and watch the ammeter.

A current flows for half of each turn but when the alternator tries to drive current the other way, none flows. The diode blocks the reverse flow and makes sure the current flows one way only. The diode changes alternating current into direct current. This process is called **rectification**.

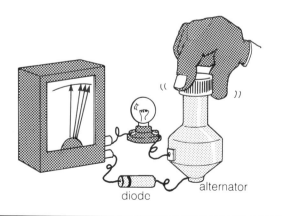 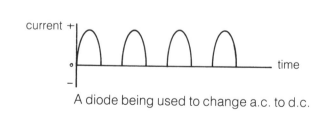

A diode being used to change a.c. to d.c.

EXPERIMENT. Looking at rectification on an oscilloscope

Connect an alternator to an oscilloscope. Turn the time base off (p. 272) and centre the spot on the screen. Rotate the alternator and watch the movement of the spot.

Repeat with a diode in series with the alternator and note the difference that it makes to the movement.

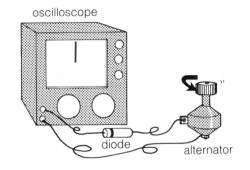

14 Copy and complete this table about the spots movement.

Description of the movement	'With' or 'without' the diode
The spot moves up from the centre line to a maximum then back down to the centre line, where it stops for half a cycle.	
The spot moves up to a maximum above the centre line and then down to a maximum below the centre line. It doesn't stop.	

EXPERIMENT. Rectifying a more rapid alternating voltage

Use a diode to rectify a small alternating voltage from a voltage box.

Connect up the circuit shown, first without a diode and then with one included. Adjust the time base of the oscilloscope until the voltage wave is steady on the screen. Draw the two voltage waves that you get with and without the diode.

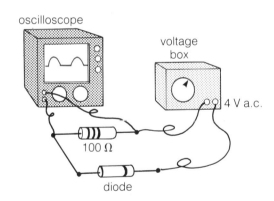

15 Which oscilloscope picture shows the following:

An alternating voltage	
A changing direct voltage	
A rectified alternating voltage	
A voltage that changes from positive to negative	
A changing voltage	
A steady voltage	

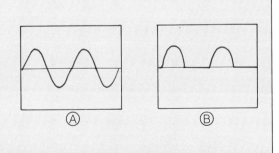

Voltage smoothing

When alternating voltage is rectified by a diode, the direct voltage does not have a constant value, and is unsuitable for use with electronic equipment. A more constant voltage can be obtained if a large capacitor (p. 181) is included in the circuit.

EXPERIMENT. The use of a smoothing capacitor

Build this simple rectifying circuit for a low-voltage a.c. supply. Connect leads from the resistor to an oscilloscope and adjust its controls to show a picture of the rectified voltage wave. Find out the effect that a capacitor has by connecting it across the resistor.

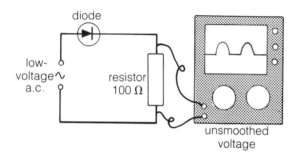

The oscilloscope picture should show that the voltage has a steadier value than before and does not dip to zero for half of the a.c. cycle.

An old radio connected in place of the resistor can also be used to indicate smoother voltage. The capacitor should make it hum less.

Do further experiments to investigate how the remaining voltage ripple depends on:
(a) the size of the capacitor,
(b) the size of the load resistor.

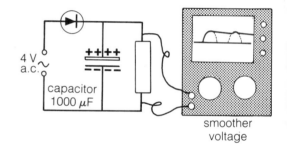

How it works

During the positive half of the cycle, the a.c. supply drives current through the resistor and into the capacitor.

During the negative half of the cycle, the diode stops current flowing from the supply. But the capacitor discharges through the resistor and maintains the voltage across it.

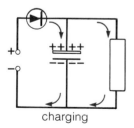
charging

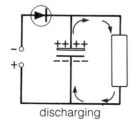
discharging

A full-wave rectifier

When a diode is used to rectify a.c., direct current flows for only one half of the alternating current cycle. No current flows during the other half. This is called **half-wave rectification**.

A better form of rectifier, called a **full-wave rectifier**, uses both halves of the a.c. cycle. It gives current in the same direction whether the alternating voltage is positive or negative. The direct current is more continuous and easier to smooth.

A full-wave rectifier can be made from four diodes as shown. Note that opposite diodes point in the same direction. The corner joining two diode points is always at a positive voltage and the corner joining two tails is always negative.

Make a full-wave rectifier and look at the rectified voltage across the load resistor on an oscilloscope.

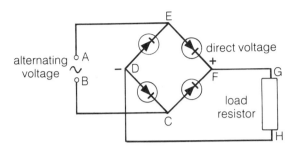

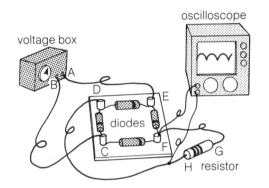

16 Use the letters on the diagram to describe the path of the current through the rectifier when:
 (a) terminal A is positive and B negative,
 (b) terminal A is negative and B positive.
 (c) What can you say about the direction of the current through the load resistor?
 (d) Where would you connect a capacitor to smooth the direct current and voltage?

Measurements with an oscilloscope

EXPERIMENT. Taking measurements with an oscilloscope

Oscilloscopes can be used to measure voltage, time and frequency.

1. Measuring the voltage of a dry cell

Set the controls as follows:
(a) time base off,
(b) input set for d.c.,
(c) Y amplifier control on 0.5 V/cm (this means it takes a voltage of 0.5 V to move the spot vertically through 1 cm).

Centre the spot, connect the cell to the input and earth, and measure the distance (in centimetres) that the spot jumps up. The voltage of the cell is given by this distance × 0.5.

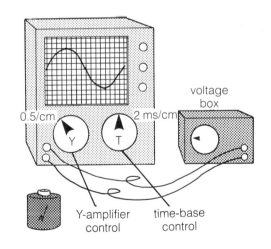

2. Measuring the peak value of an alternating voltage

Settings:
(a) time base on 2 ms/cm (this means it takes the spot 2 ms to move 1 cm across the screen),
(b) input set for a.c.,
(c) Y amplifier to 1 V/cm.

Connect a low alternating voltage to the inputs of the oscilloscope and adjust the trigger control until the wave locks. Measure the height of the wave above the centre line and multiply by 1 V/cm. This gives the peak value of the voltage. Repeat this for other voltages from the power supply and compare the values you get with its marked values.

3. Measuring the time taken for one voltage wave
Use the same settings as before but check that the time-base control is set to 'calibration'. Measure the length of one wave and multiply by 2 ms. This gives the time taken to draw one wave in milliseconds.

Calculating frequency
The number of voltage waves drawn in one second (i.e. frequency) is given by:

$$\frac{1}{\text{time taken to draw one wave}}.$$

So if the time to draw one wave is 20 ms the frequency is $1/0.020 = 50$ Hz.

17 Each picture shows a voltage trace on an oscilloscope screen. Underneath are the amplifier (A) and time base (T) control settings. For each trace calculate peak voltage, time to draw one wave and frequency.

18 Which electrical device would give which picture? Draw these oscilloscope pictures and beside each write the electrical device that produces it. (The time base is set to 'freeze' each picture on the screen.)

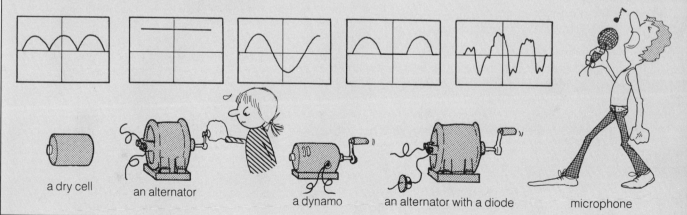

29 Electric transformers

We have seen that electricity can be made by moving a conducting wire through a magnetic field. It can also be made by moving the magnetic field rather than the conductor or changing the strength of the magnetic field.

EXPERIMENT. Generating current by moving a magnet

Connect a solenoid of many turns to a sensitive current meter. Move a magnet in and out of the solenoid and use the meter to check the following observations (tick the statements you find to be true):

A current flows whenever the magnet moves near to the solenoid.	
The current changes direction when the movement changes direction.	
There is no current when there is no movement.	
The current is larger when the movement is faster, but doesn't last so long.	

EXPERIMENT. Generating current by changing the strength of the magnetism

Use the same solenoid and meter as before but replace the magnet with an electromagnet. Fit the electromagnet inside the solenoid and connect it to a battery and switch. Use the apparatus to check the following observations (tick each statement you find to be true):

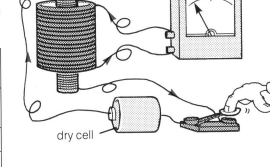

A current flows through the solenoid when the electromagnet is switched on or off.	
The current flows one way when the electromagnet is switched on and the other way when it is switched off.	
No current flows when the electromagnet is on.	
No current flows when the electromagnet is off.	
A solenoid with more turns/metre generates more current from the same change in magnetism.	

These two experiments show that current is only generated when the magnetism passing through the coil changes.

1. A magnet on a spring bounces in and out of a solenoid.
   (a) Describe the movement of the pointer of the meter.
   (b) The magnet stops bouncing sooner when it is bouncing in and out of the solenoid than when it bounces freely on the spring. Why is this? What is happening to the energy of the bouncing magnet?

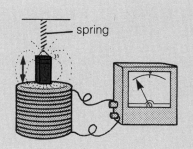

Voltage from magnetism and movement – a summary

Electricity can be forced to move along a wire by a passing magnetic field. The wire can either be moved through the field, or the field can be moved past the wire, or the strength of the field can be changed. This process is called **electromagnetic induction**.

The pictures show different ways of producing a current by electromagnetic induction. The magnetism that crosses the conductor can be made to change in three ways:
1. by moving the conductor,
2. by moving the magnetic field,
3. by changing the strength of the magnetic field without changing anything else.

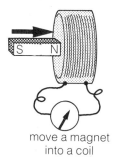
move a magnet into a coil

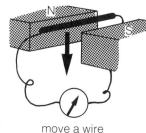
move a wire across a field

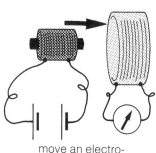
move an electro-magnet into a coil

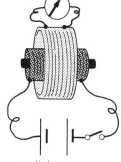
switch on an electromagnet in a coil

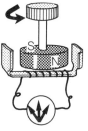
rotate a magnet near a coil

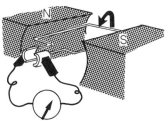
rotate a coil near a magnet

2. Decide in each case whether the magnetic field is changed by method 1, 2 or 3.

A high-voltage generator

If the solenoid in the experiment (p. 273) has a large number of turns, it generates a high voltage when the electromagnet is switched off or on. A 6 V supply to the electromagnet, for example, could generate 6000 V across the solenoid. An electromagnet inside a coil of many turns can be used to 'step-up' voltage in this way and is called an **induction coil**.

The principle is used in a petrol engine to generate a high voltage.

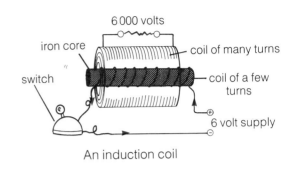

An induction coil

The induction coil in a petrol engine

The spark plug of a petrol engine needs about 15 000 V to make it spark. An induction coil is used to step-up the 12 volts of the battery to this high voltage. The induction coil has inner and outer coils, the inner one having many turns of fine wire. The current to the smaller outer coil is switched on and off by contact points. These are opened and closed by a rotating cam. The cam is driven by the engine and must open the points at the exact moment the spark is needed.

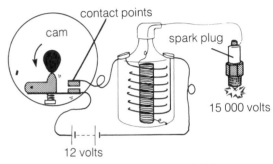

How 12 V is changed to 15 000 V in a petrol engine

3 A petrol engine with one spark plug (a single cylinder engine) has a cam shaped like (A). A petrol engine with four spark plugs (a four-cylinder engine) has a cam shaped like (B).
  (a) Can you explain why the four-cylinder cam is shaped like it is?
  (b) What would the cam from a six-cylinder look like? Try and draw it.

4 How many times does the car induction coil shown above step-up the voltage?

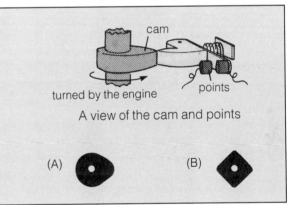

A view of the cam and points

The switch-off voltage of a relay

The coil of a relay (p. 256) is made from many turns of wire wound on an iron core. When current is switched on in the coil it strongly magnetizes the core. When the current is switched off, the magnetism collapses and generates a high back voltage across the coil. This voltage can cause sparks or damage components in the circuit (such as transistors). To remove the high switch-off voltage a diode is connected across the coil, pointing toward its positive end.

The diode 'shorts out' the high voltage but does not 'short' the normal voltage supply because this acts in the other direction.

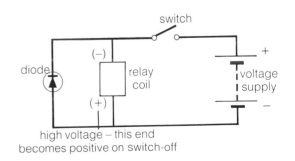

high voltage – this end becomes positive on switch-off

The transformer

A transformer, like an induction coil, is used to change voltage. It is made from two coils with different numbers of turns, wound on an iron core.

EXPERIMENT. Making a transformer

Wind about 1 m of insulated copper wire onto an iron C-core. Make another coil on a second C-core from about 2 m of the same wire and join the cores together to make a closed loop of iron. Connect a lamp to the large coil (called the 'secondary' coil) and a 2 V alternating voltage supply to the small coil (the 'primary' coil). If the lamp lights your transformer works.

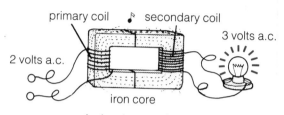

A simple transformer

Do experiments to test out the following statements (tick the ones you find to be correct):

Energy is carried from the primary coil to the secondary coil by magnetism in the core.
The brightness of the lamp decreases if you unwind some of the turns of the secondary coil. It gets brighter if you wind on more turns.
The lamp does not light if you connect 2 V d.c. to the primary coil instead of a.c. The core magnetizes but no current flows in the secondary coil.
There is a hum caused by the two C-cores attracting and releasing each other rapidly (100 Hz). This hum gets less if the two cores are held tightly together.

● EXPERIMENT

Design an experiment to see if there is a connection between the number of turns of the secondary coil and its voltage. Here is one method.

Unwind the secondary coil and use the same wire to wind on a coil of just 10 turns. Connect an a.c. voltmeter to the ends of the coil and measure the voltage generated. Wind on another 10 turns and read the secondary voltage again. Continue to add turns and read voltages until you have used all the wire.

Plot a graph of the number of turns and the secondary voltage to look for a connection.

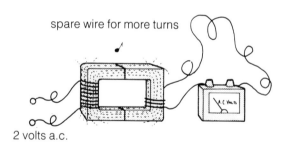

5 Copy the table and write yes or no in the boxes.

	Direct current in the primary coil	Alternating current in the primary coil
Does the core magnetize?		
Does the magnetism change?		
Does the lamp light?		

6 Why does a transformer hum?

7 Why does a transformer work without a switch?

8 What proof is there that some of the energy put into a transformer is changed into heat energy?

9 This diagram shows the energy changes that take place in a transformer. Copy it and write in these missing forms of energy: magnetic energy; heat into the coils; heat into the core.

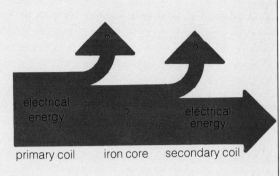

The power 'wasted' by a transformer

A transformer can be used to step down voltage. This is useful if you want to light a 12 V lamp from a 240 V 'mains' socket. To reduce the voltage like this, the secondary coil must have less turns than the primary coil (about 20 times less in this case). The diagram shows meters connected to a transformer so that the currents and voltages of each coil can be measured.

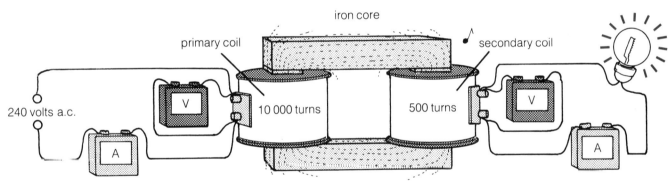

Measuring the power wasted by a transformer

Results
A very good transformer, connected as above, gave the following results:

Primary coil			Secondary coil		
Voltage	Current	Power	Voltage	Current	Power
240 V	1/10 A	12 V	12 V	2 A	24 W

These results show that, although the voltage is reduced on the secondary side, the current is increased. The results also allow us to calculate the power on each side. For a perfect transformer, the power of each coil would be the same. This shows that no energy is wasted when the voltage is stepped down. A real transformer would 'waste' energy and the coils and core would get hot.

Note also that for very efficient transformers:

$$\frac{\text{primary voltage}}{\text{secondary voltage}} = \frac{\text{number of primary turns}}{\text{number of secondary turns}}$$

---

10 These readings were obtained from not such a good transformer.

Primary coil			Secondary coil		
Voltage	Current	Power	Voltage	Current	Power
200 V	1 A		1000 V	1/10 A	

(a) Does it step the voltage up or down?
(b) Calculate the power of each coil.
(c) How much power is 'wasted' by the transformer?
(d) What happens to this wasted power?

11 Copy and complete:
'The diagram on p. 277 shows a step- _____ transformer. The secondary voltage is 20 times smaller than the primary voltage, but the secondary current is _____ times larger than the primary current.'

Energy is 'wasted' because it is converted into heat in the coils and the iron core. There is also loss of magnetic energy from the iron core. This loss is reduced by making the core 'buckle-shaped' and winding both coils on the central bar.

Transformers are used in power packs, battery chargers, voltage controllers for model trains and cars, radios, television sets and many other electronic devices. The circuit symbol for a transformer is

How electricity reaches our homes

Power stations are not usually built in towns and so the electricity they produce has to pass along wires to our homes and factories. It is best to move electricity about at extremely high voltages. Consider this example.

Transmitting power at 200 V

A small village has 'mains' electricity supplied at 200 V. At peak times, when people are using a lot of electricity, the whole village uses 200 kW of electrical power. To supply this power a current of 1000 A has to pass along the wires from the power station. This is a huge current. It would heat up the wires and cause an enormous loss of heat energy on the journey to the village.

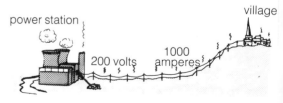

Transmitting power at high voltage and high current

Transmitting power at 100 000 V

To supply 200 kW of power at an extremely high voltage (e.g. 100 000 V) requires only a small current (2 A in this case). A small current like this would not generate much heat in thick wires and the heat loss on the journey to the village would be much less for the same power. Such high voltages would be extremely dangerous in our homes, so step-down transformers are used to reduce the voltage at the end of the journey.

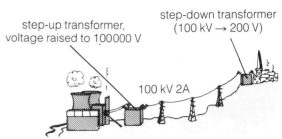

Transmitting the same power at extremely high voltage and low current

The National Grid

Electricity is carried about the country at extremely high voltages, by a network of wires (the National Grid). The Grid is supplied with electricity by a number of large power stations. When electricity is needed it is drawn from the National Grid by transformers that reduce the voltage to a level that is safe to use. Transformers play a vital part in this supply network.

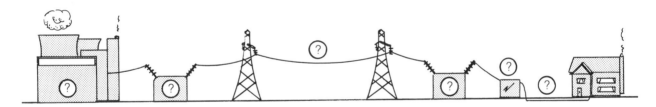

12 This diagram shows how electricity is carried from a power station to our homes. Copy the drawing and label these items: step-up transformer (11 000 V to 132 000 V); power station; step-down transformer (132 000 V to 3300 V); local step-down transformer (3300 V to 240 V); overhead high-voltage transmission lines; underground mains voltage cables.

13 Why is electrical power transmitted at extremely high voltages?

The mains voltage supply

Two wires bring electricity into the home. One wire (live) is at a high voltage and the other (neutral) is at 0 V. Appliances are connected to live and neutral and the voltage difference drives a current through them. The 'mains' electricity supply is alternating. The voltage of the live wire changes from 340 V to −340 V and back again. This causes current to flow first one way through the appliance, then the other. (This changing voltage has the same heating effect as a direct voltage of 240 V.)

The shape of the mains voltage, as it changes smoothly with time, can be seen by connecting a supply to a properly adjusted oscilloscope (do not attempt to do this). The voltage values are carefully designed to follow a sine curve that takes exactly 20 ms (1/50 s) to complete one wave.

Mains alternators have to be rotated at exactly 50 rev/s to produce these voltage waves. They generate 50 waves/s and so the alternating voltage has a frequency of 50 Hz. Mains electricity has to be made as it is used and so alternators are kept rotating night and day, at exactly mains frequency, to make it available.

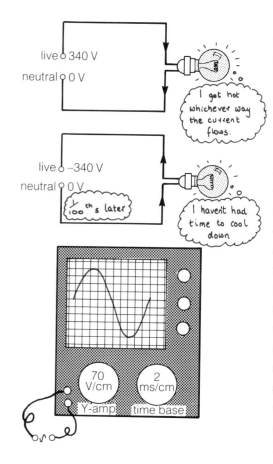

14 Explain why mains current makes a lamp shine steadily although the current through it is changing direction all the time.

15 Use the settings of the oscilloscope above and the centimetre screen grid to calculate (a) the peak voltage of the mains supply, (b) its frequency.

16 Name three appliances that can work on alternating current and two that will only work on direct current.

17 Why is it important to keep mains frequency constant at 50 Hz?

18 Two reasons that alternating (and not direct) voltage is supplied to our homes are:
 (a) it can be made in alternators without the use of brushes;
 (b) it can be transformed efficiently and so transmitted at extremely high voltage.
Explain in each case why the reason given is an advantage and why that advantage does not apply to direct current.

Electromagnetism check list

After studying the chapters on Magnetism and Electromagnetic effects you should know:
- that magnets have two poles north and south
- that a force acts through space between poles and gets less with increased separation
- that like poles repel and unlike poles attract
- that steel and alloys of cobalt and nickel are hard magnetic materials that can hold on to their magnetism
- that iron is a soft magnetic material that can be strongly magnetized but cannot hold on to its magnetism
- that the forces exerted by a magnet can be represented by a magnetic field
- that magnetic field lines show the direction and strength of the magnetic force at a place
- that magnetism can be induced in magnetic materials close to the magnet
- that a current in a long straight wire has a circular magnetic field centred on the wire
- that a current in a solenoid produces a magnetic field similar to that of a bar magnet
- that the ends of the solenoid behave like N and S poles
- that the field inside a solenoid is even and strong
- that solenoids can be used to magnetize and demagnetize pieces of steel
- that a solenoid with an iron core is an electromagnet
- that a reed switch has iron contacts that are closed by a nearby magnet
- that a relay uses a small current in a solenoid to close a switch magnetically
- that a force acts on a current in a wire that crosses a magnetic field
- that the left-hand rule can be used to find the direction of the force
- that a couple acts on a coil carrying a current in a magnetic field
- that a motor (and a loudspeaker) use this force to convert electrical energy into movement
- that a current can be generated in a wire loop by moving it through a magnetic field
- that the right-hand rule gives the direction of the induced current
- that alternating current is one that continually changes strength and direction
- that rectification means changing a.c. to d.c.
- that a diode can be used to half-rectify alternating current
- that a capacitor can be used to smooth the output of a rectifier
- that an oscilloscope can measure voltage from the deflection on its screen and the Y-amplifier setting
- that an oscilloscope can measure time from the length of a wave on the screen and the time-base setting
- that a transformer can step-up or step-down voltage and current
- that the primary/secondary voltage ratio depends on the primary/secondary turns ratio
- that electrical energy is generated and transmitted at very high voltages, to reduce heating of the wires
- that transformers play an essential role in raising and lowering these voltages
- that mains electricity is a.c. with a frequency of 50 Hz and a voltage equivalent to 240 V d.c.

Revision quiz

Use these questions to help you revise the work on Magnetism and Electromagnetic effects.

- How can you find the N pole of a magnet?

    ... by noting which end points to the Arctic when the magnet is freely suspended.

- What is the law of force between poles?

    ... like poles repel, unlike poles attract.

- What is a hard ferromagnetic material?

    ... ones such as steel and alloys of nickel and cobalt that hold onto their magnetism, once magnetized.

- What is a soft ferromagnetic material?

    ... one such as iron that can be magnetized strongly but that loses its magnetism as soon as the magnetizing force is removed.

- What is a magnetic field line?

    ... a line in a magnetic field that shows the direction of the force on a N pole.

- What is magnetic induction?

    ... the creation of N and S poles on pieces of iron and steel placed in a magnetic field.

- Why does a magnet attract pieces of iron?

    ... poles are induced on the iron and the nearer one is attracted to the magnet with a greater force than the repulsion on the further one.

- What is the shape of the magnetic field of a current in a long straight wire?

    ... a field that is circular in cross-section, centred on the wire.

- What is the field like inside a solenoid carrying a current?

    ... uniform and relatively strong.

- When are relays used?

    ... when we want a small current from a device like a transistor to switch on a large current for a motor or heater.

- What is the 'motor' force?

    ... the force on a current as it passes across a magnetic field. The force, the field and the movement are all at right angles.

- How can current be generated mechanically?

    ... by moving part of a closed loop of wire through a magnetic field.

- How can the size of this current be increased?

    ... by increasing the number of turns of the wire, by moving the wire faster and by increasing the strength of the magnetic field.

- Which rules give the directions of the motor force and the induced current?

    ... left-hand rule for motor force and right-hand rule for the current direction.

- Describe alternating current.

    ... it flows first one way, increasing then decreasing in strength, then the other.

- What is rectification?

    ... changing a.c. to d.c.

- How can an oscilloscope measure voltage?

    ... by multiplying the deflection of the spot on the screen (in centimetres) by the setting of the Y-amplifier knob.

- How can a transformer be made to step up voltage?

    ... by having a secondary coil with more turns than the primary coil.

- What is the ratio equation for a perfect transformer?

    $V_p/V_s = N_p/N_s$

- How is power lost in a real transformer?

    ... by heat in the wires, heat in the core and loss of magnetism from the core.

- Why does the National Grid distribute mains electricity at very high voltages?

    ... so that smaller currents can be used and heating losses are less.

Examination questions

1. (a) A coil of wire is connected to a battery and a switch as shown in the diagram.
   (i) Draw on the diagram the shape of the magnetic field in and surrounding the coil when the switch is closed. Clearly mark the current direction, and the field direction.
   (ii) How would this arrangement have to be changed to form an electromagnet?
   (iii) Name one practical device which uses an electromagnet.

   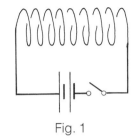
   Fig. 1

   (b) Two metal rods are placed inside a coil as shown. When direct current flows in the coil, the rods move apart, but when the current is switched off, they return to their original positions.
   (i) Why do the rods move apart?
   (ii) From what metal could the rods have been made? Explain your choice.
   (iii) If alternating current from a mains transformer was used instead of direct current, what effect (if any) would this have on the rods? Explain your answer.

   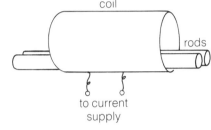
   Fig. 2

   (c) The diagram shows a heavy copper rod freely suspended on thin bare copper wires between the poles of a magnet. The ends of the suspension wires are connected to the input of an oscilloscope which is suitably adjusted with its time-base on.
   (i) Explain why a voltage is induced across the rod when it is swinging between the poles of the magnet.
   (ii) What type of voltage is induced?
   (iii) Sketch the appearance of the trace on the oscilloscope screen when the rod is swinging freely.
   (iv) If the length of the suspension wires is adjusted so that the rod is made to swing faster and at a higher frequency, what TWO changes would there now be in the appearance of the trace? (*NISEC*)

   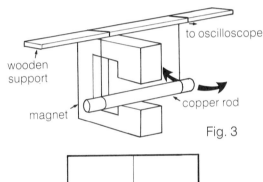
   Fig. 3

   Fig. 4

2. Fig. 5 shows the screen, Y-gain and time-base controls from a typical oscilloscope displaying a waveform.
   (a) What is the setting of the Y-gain control?
   (b) What is the peak voltage of the waveform?
   (c) What is the time-base setting?
   (d) What is the period of the trace?
   (e) What is the frequency of the waveform?
   (f) If the time-base setting is altered to 1 ms cm^{-1} and the Y-gain to 2 V cm^{-1} copy and draw the resultant trace on the graticule shown below. (*NEA*)

   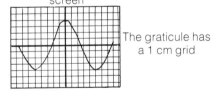

   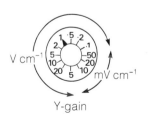 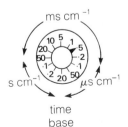
   Fig. 5

   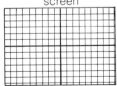
   Fig. 6

3 (a) An electromagnet is made by winding a coil of wire on a soft iron rod.
   (i) When a current flows in the coil, what happens to the soft iron rod?
   (ii) Why is hard steel not suitable for use as the core of an electromagnet?
 (b) The next diagram shows an electromagnetic relay, which is a type of switch using an electromagnet.
   Explain why the lamp lights when the switch S is closed.
   (LEAG)

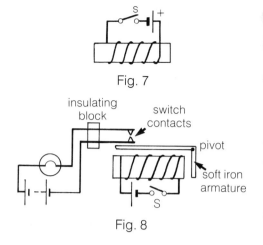

Fig. 7

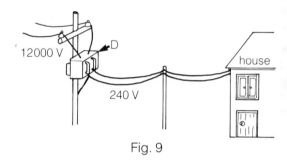

Fig. 8

4 This question is about supplying a consumer with electrical power from the National Grid system.
 The voltage across the power lines supplying alternating current to an isolated house is 12 000 V.
 The device D changes the voltage of the supply to 240 V.
 (a) What do we call the device D?
 (b) Why is the supply not transmitted all the way at 240 V?
 (c) Why cannot the 12 000 V supply be used, unchanged, in the house? Give TWO reasons.
 (d) Why is alternating current used? *(SEG)*

Fig. 9

5 The circuits show two ways of switching an electric motor on and off. Circuit 1 uses a simple switch. Circuit 2 uses a reed relay as a switch.
 (a) Explain the working of the reed relay.
 (b) Describe a situation where circuit 2 is better than circuit 1 for controlling a motor. *(MEG)*

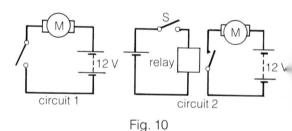

Fig. 10

6 The coil of a reed relay is connected to a 6 V alternating current supply as shown above. The frequency of this supply is 4 Hz. Describe what happens in the light-emitting diode circuit when this supply is switched on. *(MEG)*

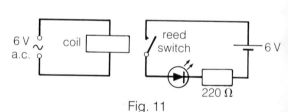

Fig. 11

7 (a) (i) Draw a labelled diagram of a transformer suitable for converting the 240 a.c. mains to 12 V a.c.
   (ii) If the transformer is ideal and has 80 turns on its secondary (12 V) winding, how many turns should it have on the primary winding?
 (b) Explain how the transformer works.
 (c) Transformers are usually not 100% efficient. Give a full explanation of two ways in which energy losses occur.
 (d) Explain the part played by transformers in the distribution of electrical energy around the country. *(MEG)*

30 Electronics – the transistor

Transistors are components that have many uses as electronic building blocks. They can be used as fast electronic switches that have no moving parts, and as amplifiers.

Transistors all have three working leads that are named:
collector
base
emitter

The circuit symbol shows the three leads and the arrow inside shows whether the transistor is an 'npn' or 'pnp' type.

Current flows into a npn transistor through two entrances (collector and base) and out by one exit (emitter). The current that goes into the collector entrance is called the collector current. The current that goes into the base entrance is called the base current. The circuit diagram shows how the two currents can be supplied from a single battery. The circuit can also be used as a transistor tester.

The two types of transistor

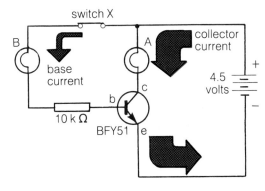

PROJECT. Building a transistor test circuit

One way to make this circuit is with terminal pins on matrix board. This is an insulating board covered with holes that are just the right size for the terminal pins.

Draw out the circuit on the board and put in pins at all the corners. Include a triad of pins for the legs of the transistor. Use thin copper wire to connect one pin to another where there are links in the circuit. Twist the wire round each pin and solder it in place. Then solder the components into their places. The switch is just a wire with a loose end that can bridge the gap between two pins. It switches the base current on and off.

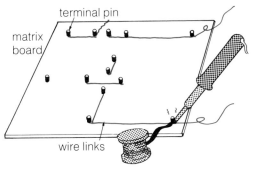

The transistor is working if there is:
1. no collector current (lamp A off) when there is no base current (switch open);
2. collector current flows (lamp A on) when there is a base current (switch closed).

Your circuit will also allow you to examine these statements about transistor currents:

1 The collector current cannot flow unless the base current flows.

2 The base current is much smaller than the collector current. The base current passes through its lamp (prove this by unscrewing the lamp) but is too small to light it.

A transistor uses the idea of 'small' controlling 'large'. The small base current controls the much larger collector current.

The same idea is used in a car engine. A small force on the accelerator controls the large force of the engine. A device that can use small changes to produce much bigger effects has many applications.

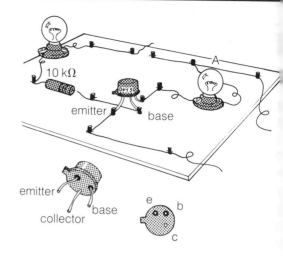

1 Explain these examples of 'small' controlling 'large': accelerator and engine, rudder and ship, the little men in the picture, aileron and plane, base current and collector current.

A few controlling the many

EXPERIMENT. The transistor as a switch

Use this circuit to find the base voltage that is needed to turn on the collector current.

Build it on matrix board with the components shown and a digital voltmeter (or one with a high resistance) between the base pin and the − pin. Use the potentiometer to gradually turn up the base voltage from 0 V until the lamp suddenly comes on. Measure the voltage for different transistors if you can.

Most transistors 'turn on' when the base voltage is about 0.6 V or higher. Do yours behave this way?

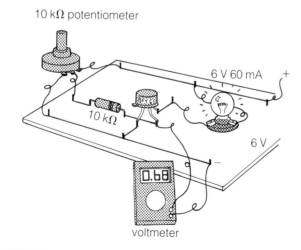

PROJECT. A simple twilight switch

The lamp in this circuit comes on automatically when it gets dark. The circuit uses a photocell (ORP 12) that has a low resistance in the light and a high resistance in the dark. As it gets dark, the resistance of the ORP12 increases and pushes up the voltage of the base pin. When this reaches 0.6 V, collector current starts to flow and the lamp lights.

Build the circuit on matrix board and check that it works. Then use a digital voltmeter to watch how the base voltage changes with the light level.

Does it behave as theory says it should?

A twilight switch

2 Suggest some uses for this electronic circuit.

PROJECT. Not-so-simple twilight switches

The following circuits show improvements you can make to the basic light operated switch.

The light level falls .... the resistance of ORP12 rises .... the base current increases .... the collector current increases ... the lamp goes on.

1. Light-level control

The 100 kΩ potentiometer can be used to make the transistor switch on at different light levels. To adjust it for a certain light level, turn the potentiometer until the lamp comes on and then turn it back until the lamp just goes off. A small decrease in brightness should then trigger the switch.

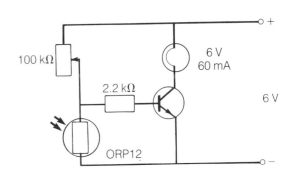

2. Increased sensitivity

The two transistors work together to make the circuit switch on more suddenly. The pair (known as a Darlington pair) have a collector, base and emitter and can be looked on as a single transistor.

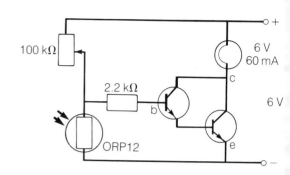

3. Switching large currents

Transistors cannot pass enough current to drive heaters, large lamps or motors.

If the lamp in the circuit is replaced by the coil of a relay (p. 256), 'darkness' will magnetize the relay and close its contacts. These contacts can then switch on currents in a second circuit that drive the motors or large lamps.

The diode is needed to short out the large 'back' voltage that is generated when the relay coil demagnetizes. Without it the transistor would be damaged (p. 275).

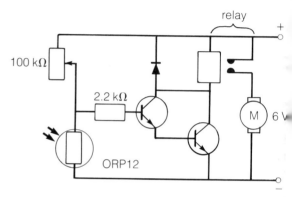

4. Switching on in the light

If the light-dependent resistor and potentiometer are changed over, the transistor will switch on in the light. (Use a 5 kΩ potentiometer instead of 100 kΩ.)

Use this idea to build a circuit that uses a torch beam to switch on an electric motor.

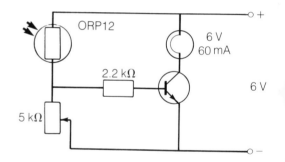

31 Electronics – microchips that make decisions

Complete electronic circuits can now be built on tiny pieces of silicon and sealed into small packages or 'microchips'. One of the many different types of microchip is the 'logic chip'. This can be used to build circuits that make decisions.

Logic chips have names like NOT, AND and OR that describe the type of logical decision they make.

The NOT gate

Suppose we want an electronic device that can 'decide' when the water level in a kettle is too low and tell us by switching on a warning light. We can use a logic chip called a NOT gate and two simple wire probes. The chip is arranged to show whether or NOT the probes are under water.

A NOT gate has one input pin and one output pin. When its input voltage is high, then its output voltage is low and vice versa ('high' means about 5 V and 'low' about 0 V). It has a high output when its input is NOT high.

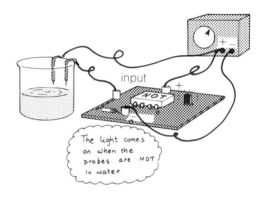

The light comes on when the probes are NOT in water

Circuit design

Work with logic chips is most easily done on ready made 'kit' boards with plug-in outputs and inputs. The chips themselves must have a +5 V and 0 V voltage supply and usually have an LED to indicate whether the output voltage is high or low. The same voltage supply can be used for the chip's inputs.

The probes for the water level warning device can be made from two pieces of stiff wire. One should be connected to the + of the voltage supply and the other to the input of the NOT gate. You will then find that when both probes are in water the LED is out but when one probe is NOT in water the warning light comes on. The LED comes on when the probes are in air because the input voltage is NOT high. Build a model of this warning device to find out how well it works.

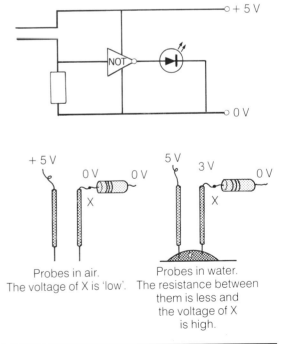

Probes in air. The voltage of X is 'low'.

Probes in water. The resistance between them is less and the voltage of X is high.

The truth table of a NOT gate

A truth table (p. 201) shows the gate's output voltages for high and low input voltages. High voltage is around 5 V and is called level '1'. Low voltage is around 0 V and is called level '0'. Check this truth table of a NOT gate by connecting high and low voltages to its inputs. Note the circuit symbol for the NOT gate.

INPUT	OUTPUT
0	1
1	0

A NOT gate

Other NOT projects to try

1. The same circuit can be used to indicate when a cup or beaker is NOT level. If the probes are fixed to the opposite sides of the cup, any tilt will expose one probe and the warning light will come on.

2. Bath overflow indicator. If you connect a second NOT gate to the output of the first one, its output lamp will light when the probes touch water. The probes could then be fixed near to the top of the bath where they will give a warning when the water reaches them.

The OR gate

Suppose we want to make a wet-nappy detector for twins! This would give a warning if one twin OR the other has a wet nappy. The logic chip that can make this decision is an OR gate.

It has two inputs and one output. If one of the inputs OR the other is made high, the output goes high and can be used to trigger an alarm.

Circuit design

Make two dampness probes from stiff wire and connect them to the inputs of the OR gate as shown. Push the probes into pieces of nappy cloth or tissue. A drop of water on either nappy should then trigger the alarm light.

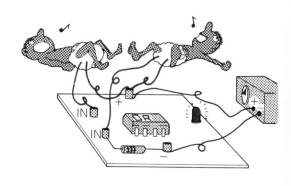

The truth table of an OR gate

Check the truth table for an OR gate by making each input high in turn and using an LED to show whether the output is high or low.

A	B	OUTPUT
0	0	0
0	1	1
1	0	1
1	1	1

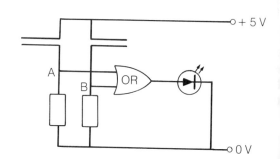

Symbol and truth table for an OR gate

Another OR project to try

You may like to design and build a circuit for a 'dry plant detector'.

This could have probes at the top and bottom of a plant pot that detect when the earth is dry at one level OR the other. The detector could have a green OK light that goes out when water is needed. Or, by adding a NOT gate, could have a red warning light that goes on when water is needed.

The AND gate

Suppose we want a cooling fan in a room that comes on automatically when it is too hot but only if the lights are on and there are people present. The logic chip needed is an AND gate. This will only switch on the fan if it is too hot AND the lights are on.

To sense light a voltage-divider is needed, made from an LDR and a potentiometer. The temperature sensor is made from a thermistor and another potentiometer. The potentiometers make it possible to vary the levels of temperature and light at which the fan comes on (p. 225).

The gate cannot supply enough current to drive a fan motor, so a transistor must be connected to its output to switch on the motor.

The complete circuit looks like this; build it and see how well it works.

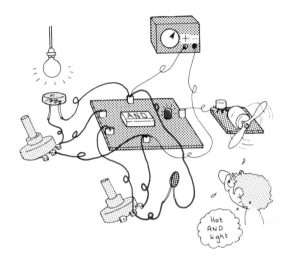

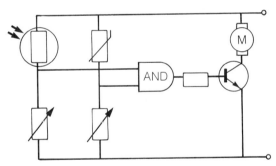

A	B	OUT
0	0	0
0	1	0
1	0	0
1	1	1

Symbol and truth table for an AND gate

Negative gates

Two other useful logic microchips are the NAND (Negative AND) and NOR (Negative OR) gates. These work in the opposite way to AND and OR gates as shown by their truth tables. Confirm the truth tables for yourself if you can and think out some uses for negative gates.

A	B	OUT
0	0	1
0	1	1
1	0	1
1	1	0

NAND

A	B	OUT
0	0	1
0	1	0
1	0	0
1	1	0

NOR

1 Copy this summary table and put AND, NAND, OR, NOR, NOT alongside the correct description of its logic.

If either input is high the output is high.	
Only if both inputs are high is the output high.	
If the input is high the output is low.	
Only if both inputs are low is the output high.	
If either input is high the output is low.	
If the input is low the output is high.	

2 Copy this summary of the truth tables of four logic gates. Write the correct gate and circuit symbol in the blank spaces.

A	B					Name
0	0	0	1	0	1	
0	1	1	0	0	1	
1	0	1	0	0	1	
1	1	1	0	1	0	
						Circuit symbol

British Standards circuit symbols for logic gates (AND, OR, NOT, NAND, NOR)

Combining logic gates

Logic gates can be combined together to make circuits that meet special needs. Here are two examples:

1. Two NOR gates connected to make a 'latch'.
If input B is held low and input A goes high for a moment, output C goes high and stays high even if A goes low again. This latching type of circuit is useful in burglar alarms, where a brief signal sets off an alarm that continues to ring even after the signal has stopped.

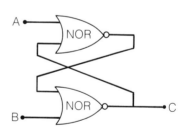

2. Combining gates for a special job.
This example shows how logic gates can be combined to give a special truth table that solves a particular problem. Check the truth table for yourself by working through each gate in turn.

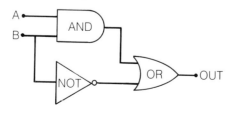

A	B	OUT
0	0	1
0	1	0
1	0	1
1	1	1

Electronics check list

After studying the chapters on Electronics you should know:
- that a transistor has three working legs called collector, emitter and base
- that the collector and base are current entrances and the emitter is an exit
- that a much larger current enters the collector than the base
- that the small base current controls the collector current
- that a transistor can be used as a fast acting switch with no moving parts
- that a small base current (or base voltage > 0.6 V) will switch on the collector current
- know how to use a thermistor or LDR to switch on a transistor
- that logic microchips work on a voltage of 5 V (called logic '1') and 0 V (called logic '0')
- that if either input of an OR gate is high, the output is high
- that if either input of a NOR gate is high, the output is low
- that if both inputs of an AND gate are high, the output is high
- that if both inputs of a NAND gate are high, the output is low
- that if the input of a NOT gate is high, the output is low and vice versa
- know the truth table of each of these gates
- know examples of devices that use logic gates

Revision quiz

Use these questions to help you revise the work on Electronics.

- Which legs of a transistor act as current entrances?
- How are the collector and base current connected?
- What base voltage is needed to switch on the collector current?
- What is the purpose of the base resistor?
- How can an LDR be used to switch on a transistor?
- How can a thermistor be used to switch on a transistor?
- How can the collector current be made to work large current devices?
- What is a microchip?
- What is a logic chip or logic gate?
- What is meant by logic '1' and logic '0'?
- What is the logic of a NOT gate?
- What is the logic of an OR gate?
- What is the logic of a NOR gate?
- What is the logic of an AND gate?
- What is the logic of a NAND gate?
- How can logic gates be used?

... currents flow into the base and collector, and leave by the emitter.

... the base current is about 100 times smaller than the collector current, yet has control over it.

... 0.6 V or above.

... to prevent the base current from becoming too large and damaging the transistor.

... make it part of a voltage divider by connecting it with a potentiometer across a voltage supply. Then connect the junction of the two to the base resistor.

... make it part of a voltage divider by connecting it with a potentiometer across a voltage supply. Then connect the junction of the two to the base resistor.

... by passing it through the solenoid of a relay and using the heavy duty relay contacts to switch on the device.

... a complete circuit built on a piece of silicon and sealed into a small package.

... a microchip that has been designed to give an output voltage for a set pattern of input voltages.

... a 'high' voltage around 5 V is called logic 1 and a low voltage around 0 V is called logic 0.

... it inverts its input voltage so that a high input voltage gives a low output and vice versa.

... it gives a high output if either one OR the other of its inputs is high.

... it gives a LOW output if either one or the other input is high.

... it gives a high output if one of its inputs AND the other is high.

... it gives a LOW output if one of its inputs AND the other is high.

... they can sense voltages from probes and give outputs based on 'logical decisions'.

Examination questions

1. In this circuit the relay coil is energized and can close switch S if the *output* of the NOT gate is high.
   (a) When a bright light is shone on the light-dependent resistor the *input* voltage to the NOT gate falls.
      (i) Why does this happen?
      (ii) What happens in the rest of the circuit as a result of this?
   (b) Suggest a practical use for the circuit shown above.
   (*MEG*)

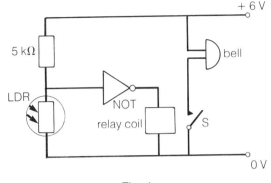

Fig. 1

2. Fig. 2 shows a relay being driven by a transistor. The diagram also shows a 12 V power supply and electric immersion heater connected to the relay contacts.
   (a) A component is usually placed between X and Y to protect the transistor.
      (i) What is the name of this component?
      (ii) Copy and draw in the component between X and Y.
   (b) The relay is of the 'normally-open' type, i.e. the contacts are open when the relay coil has no current flowing through it. What is the state of the contacts when the voltage between M and N is (i) 0 V, (ii) +5 V?
   (c) (i) The circuit in Fig. 2 can be made to operate as a thermostatically controlled heating circuit by the addition of a variable resistor and another component at the input to the transistor. Copy and draw in these two components, correctly connected, so that the circuit will work.
      (ii) Which component can adjust the temperature at which the relay operates?
   (*NCA*)

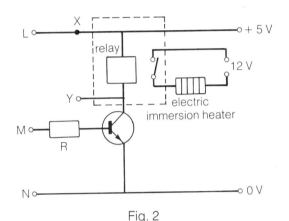

Fig. 2

3. (a) A NOT logic gate can be added to the output of a two-input OR logic gate to produce a two-input NOR (NOT OR) logic gate. The completed truth tables for a NOT and an OR gate are given on the right.
   Copy and complete the truth table for a two-input NOR gate.

A	B	P
0	0	
0	1	
1	0	
1	1	

   NOT

A	P
0	1
1	0

   OR

A	B	P
0	0	0
0	1	1
1	0	1
1	1	1

   A and B are inputs. P is the output.

   (b) The circuit symbol for a two-input NAND (NOT AND) gate is shown on the right. Write down its truth table.

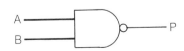

Also given is the truth table for a two-input NAND gate. In the design of logic circuits, the NAND gate is a basic 'building block' as in the following circuit.

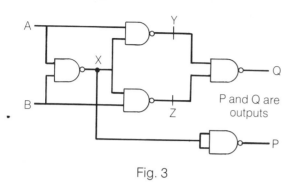

A	B	P
0	0	1
0	1	1
1	0	1
1	1	0

Fig. 3

(i) Copy and complete the truth table for this circuit.

A	B	X	Y	Z	P	Q
0	0	1				
0	1	1				
1	0	1				
1	1	0				

(ii) What arithmetic operation does this circuit perform?

(c) The output of a logic circuit can be displayed using an LED and associated series resistor, as in the circuit shown below.
(i) When the LED is lit, what is the logic state at A?
(ii) The LED has a potential difference drop of 2 V across it and a current of 10 mA flowing through it when it is lit. What is the potential difference across resistor R when the LED is lit?
(iii) Calculate the resistance of R.
(iv) Why is the resistor R needed?

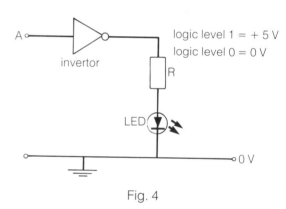

logic level 1 = + 5 V
logic level 0 = 0 V

Fig. 4

(d) In a factory, a particular piece of machinery has an alarm system to warn the operator of a fault. The system has a blue lamp (alight when the machine is operating normally), a red warning lamp and a buzzer. There is a fault detection system and an 'operator acknowledge' button. If a fault is detected, the blue lamp goes out and the red lamp comes on and the buzzer sounds. When the operator acknowledges the fault by pressing the button, the buzzer stops but the red light stays on. Complete the output columns of the truth table below to show what states are wanted.

(NEA)

Input		Output		
Fault	Operator acknowledge	Red	Buzzer	Blue
0	0			
0	1			
1	0			
1	1			

Section 3
MATTER

32 Measuring matter

Mass

The mass of an object measures the amount of matter in it. Mass is measured in kilograms (kg) or sometimes grams (g). 1000 g = 1 kg. When we buy 1 kg of fruit we expect to get a certain amount. The shopkeeper measures this amount by balancing it on a scale against a 1 kg mass of metal. Modern balances give a direct reading but have to be checked with 'standard' 1 kg masses. These standard masses are the same all over the world.

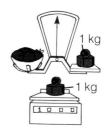

EXPERIMENT. Measuring mass

Learn to use the balances you have in the laboratory. Find the masses of some of the things you have in your pockets.

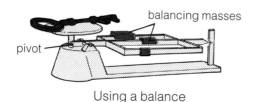

Using a balance

1 What are the masses of these objects? Choose from these values: 20 g; 110 g; 1 kg; 4 kg; 25 kg.

A sack of cement	A baby	A jar of coffee	A loaf of bread	An airmail letter

Volume

A cube with 1 centimetre sides has a volume of 1 cubic centimetre (1 cm^3). The volume of an object is the number of these centimetre cubes it contains. The block shown fills the same space as 24 of these cubes, so its volume is 24 cubic centimetres. Length × breadth × height (4 × 3 × 2 cm^3) also gives the volume.

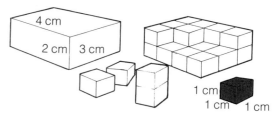

EXPERIMENT. Using a measuring cylinder to measure volume

Measuring cylinders are marked in cubic centimetres and so can be used to measure the volume of objects. Use a regular block and find its volume from change in water levels. Calculate its volume by measuring its sides and multiplying length, breadth and height. Are the two values the same?

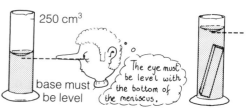

Using a measuring cylinder

2. How would you use a measuring cylinder to measure the volume of an object (a) that sinks, (b) that floats.

3. What happens to the water in the displacement can when the key is dropped in? How could you measure the volume of an object that will not go into the measuring cylinder?

4. Volume can be measured in cubic metres (m³). How many cubic centimetre blocks are there in a cube with 1 m sides?

5. How would you find the volume of liquid held by a teaspoon.

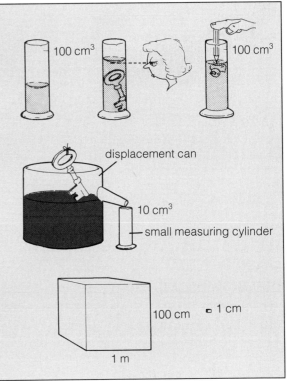

Density

We must be careful before we say that steel is heavier than wood because it depends on the volume we are talking about. To compare the heaviness of materials we must take the same volume of each. A cube of steel has much more mass than a cube of wood with the same volume.

EXPERIMENT

Measure the masses of blocks of different materials. Use blocks that all have the same volume. Some materials have more mass per cubic centimetre than others. They are more **dense**. The **density** of a material is its mass per cubic centimetre. List your materials in order of their densities.

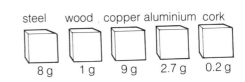

6. Put the materials shown above in order of their densities, the most dense first.

7 A block of material measures 5 cm × 3 cm × 2 cm and has a mass of 270 g.
   (a) How many 1 cm³ blocks would fill the same volume?
   (b) What would be the mass of one of these blocks?
   (c) What do you think the block is made of? (Look at the density figures in question 6.)

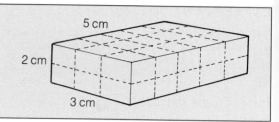

The density of any substance of any shape can be found by dividing its mass by its volume. This will give the mass of 1 cm³.

$$\text{density} = \frac{\text{mass}}{\text{volume}}$$

The units commonly used are grams/cubic centimetre (g/cm³) or kilograms/cubic metre (kg/m³).

8 Two objects are balanced on a simple beam balance. Say whether the following quantities are the same or different for the objects: mass, volume, density.

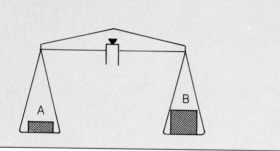

EXPERIMENT

Find the density of some common materials such as steel, glass, wood and aluminium. Find their volumes by measuring (length × breadth × height) or by using a measuring cylinder. Calculate density by **dividing mass by volume.**

9 Copy this table and work out the numbers for the empty spaces.

Material	Mass (g)	Volume (cm³)	Density (g/cm³)
Steel	160	20	
Wood	150	150	
Lead	110		11
Glass		50	2

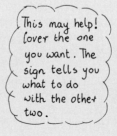
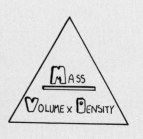

EXPERIMENT. Measuring the densities of some liquids

Find the mass of 100 cm³ of a liquid in a measuring cylinder. You will have to find the mass of the cylinder empty and then with the liquid in it. Find the densities of water, oil, paraffin, methylated spirits if you can.

Example of results
Mass of the measuring cylinder = 200 g.
Mass of the measuring cylinder with liquid = 280 g.
Mass of the liquid = 80 g.
Volume of the liquid = 100 cm³.
Density of the liquid = 80/100 = 0.8 g/cm³.

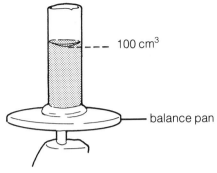

Measuring the density of a liquid

The density of air

Has air got mass? A sensitive balance shows that a bottle has a little more mass when full of air than when the air has been taken out.

EXPERIMENT. Measuring the mass of 1 litre (1000 cm³) of air

Connect a 1 litre flask to a vacuum pump by a rubber tube that can be closed by a screw clip. Use the vacuum pump to remove the air from the flask. Then close the clip. Find the mass of the flask and its fittings by placing them on a sensitive balance. Open the clip and air will rush in and fill the flask. The balance reading should rise a little showing that air has mass. Work out the mass of this litre of air from the balance readings.

Results – using a 1 litre flask.
Mass of the flask and its fittings when empty of air = 340 g.
Mass of the flask and its fittings when full of air = 341.2 g.
Mass of air = 1.2 g.

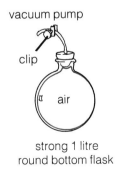

Measuring the density of air

10 The flask from the experiment above is emptied of most of its air and its tube is placed under water.
Explain:
(a) what will happen when the clip is opened,
(b) how the water can show whether the flask was completely empty of air.

11 If a litre of air has a mass of 1.2 g, what will be the mass of 1 cubic metre of air? (1000 litres = 1 cubic metre.)

12 A room measures 3 m × 4 m × 5 m. Calculate:
(a) the volume of the room in cubic metres,
(b) the mass of air in the room if the density of air = 1.2 kg/m^3.

13 Copy and complete this table.

Object	Mass (g)	Volume (cm^3)	Density (g/cm^3)
A	16	2	
B	8	4	
C	4		8
D	8	16	

(a) Which object has the greatest mass?
(b) Which object has the least volume?
(c) Which objects could be made of the same material?
(d) Which object would float on water?
(Take the density of water as 1 g/cm^3.)

14 The following liquids and solids were put into a measuring cylinder and allowed to settle.

Oil	(0.8 g/cm^3)
Water	(1.0 g/cm^3)
Cork	(0.2 g/cm^3)
Polythene	(0.9 g/cm^3)
Mercury	(13 g/cm^3)
Steel	(8 g/cm^3)

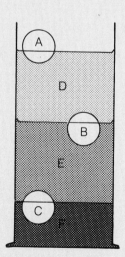

Copy and put in the letter that shows the substance on the diagram.

33 The change of state

Heat can make materials change state

Set up these heating experiments and watch what happens.

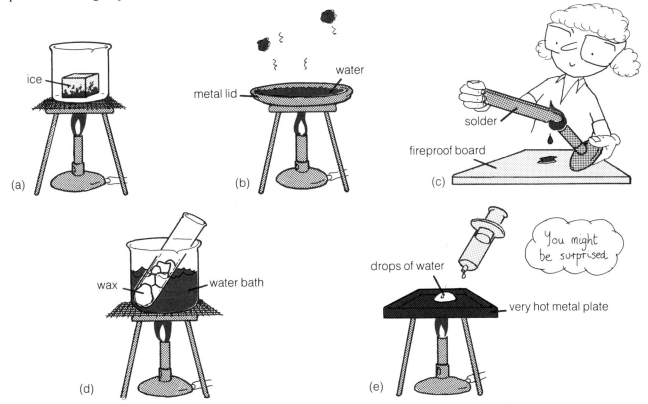

1. What will happen in these experiments? Copy the table and fill in the spaces.

Material that is heated	Is it solid, liquid or gas at first?	Does it boil?	Does it melt?	What does it turn into?
Ice				
Water				
Solder				
Wax				
Drops of water				

Materials are either solids, liquids or gases. These are the three states of matter. Heat can change matter from solid to liquid and from liquid to gas.

> **2** What must be done to change a gas into a liquid or a liquid into a solid?

Boiling

We often boil water. Heat warms the water to its boiling point and then changes it from liquid to gas.

⬤ **EXPERIMENT. Watching water boil**

Heat a beaker of fresh tap water and watch it carefully until it boils. Put a thermometer in the water and write down the temperatures when you see the following things happen:

(a) Mist appears on the outside of the beaker.
(b) Small bubbles rise from the bottom of the beaker to the top.
(c) Wisps appear above the surface.
(d) Large bubbles appear and disappear on the bottom of the beaker.
(e) The large bubbles get to the surface where they burst, whipping the surface into a turmoil.

It's surprising how much happens

Boiling occurs when large bubbles form in the liquid, then rise to the surface and burst.

> **3** Water and alcohol were heated separately and the temperatures of the liquids were taken every minute.
>
> | Time (minutes) | 0 | 1 | 2 | 3 | 5 | 8 | 11 | 14 |
> |---|---|---|---|---|---|---|---|---|
> | Water temp. (°C) | 20 | 27 | 35 | 42 | 61 | 83 | 100 | 100 |
> | Alcohol temp. (°C) | 20 | 40 | 57 | 70 | 78 | 78 | 78 | 78 |
>
> Plot a graph of temperature against time for each liquid.
> (a) Explain why the graphs level off.
> (b) What are the boiling temperatures of water and alcohol?

Hidden heat

If you continue to heat water when it is boiling, its temperature does not rise. So what is the heat energy doing? Heat energy changes the boiling water from a liquid to a gas. This energy is held by the steam and is released when the steam turns back into water. Steam at 100 °C has more energy than the same amount of water at 100 °C. The extra energy needed to change water into steam is called **latent** (hidden) heat.

the steam carries energy

> **4** If steam from a kettle is allowed to condense on a spoon, the spoon gets hot. Explain how the heat is carried from the water to the spoon.

Evaporation

Wet clothes dry in the wind and puddles of water disappear. What happens to the water? The liquid water must turn into an invisible vapour that is carried away by the wind.

The more energetic water molecules manage to escape from the liquid and move around among the air molecules. This happens even if the water is not boiling.

When a liquid turns into a gas without boiling, it is said to be evaporating. It is hard to detect water vapour but when scent or aftershave evaporate we can smell the vapour.

> 5 Does water have to be at its boiling point before it can evaporate? Give an example that proves your answer.

EXPERIMENT. Evaporation

Dip a finger into acetone. Lift it out and watch the acetone evaporate. The liquid disappears but do any of the molecules reach you nose? Notice that your finger feels a little cooler as the liquid evaporates.

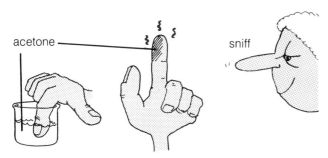

EXPERIMENT. Cooling by evaporation

Dip a thermometer into acetone and read the temperature. Remove the thermometer and read the temperature as the acetone evaporates from the bulb. Find the drop in temperature caused by the evaporation of other liquids. Try water and alcohol for example.

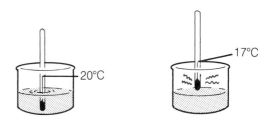

Why does evaporation cause cooling?

Gas molecules have more energy than the molecules in a liquid. A liquid cannot turn into a gas unless it gets energy from somewhere. If the energy is not supplied by a heater, then the liquid will take the energy it needs from the things around, cooling them down.

> 6 Explain why you feel cold when you get out of a pool or bath, especially if it is draughty. Note – the water itself is not cold. Why do you feel cold for longer if you are wearing wet clothes?
>
> 7 How does perspiration keep us cool?

Condensation – the reverse of evaporation

Clouds

If water vapour enters cool air then the vapour turns back into liquid, not as a puddle but as many tiny drops of pure water. Clouds are made up of millions of water drops held up in the air.

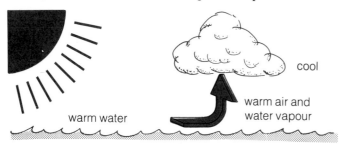

Dew

During the day the warm air collects water vapour. In the night the air cools down. Vapour condenses and forms drops of water on the ground. Dew drops like to form on sharp points or edges.

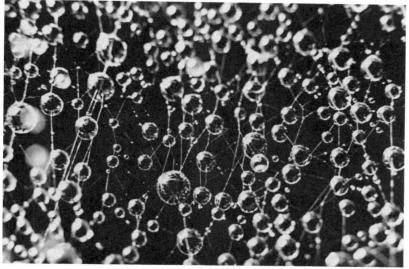

Dew drops on a spider's web

---

8 Explain why a person's glasses mist up when he walks into a steamy bathroom from a cold passage. Or when he sips hot soup.

9 Copy and complete.

	Evaporating	Condensing
What change of state takes place?		
Does energy have to be supplied or removed?		
Are the particles moving further apart or closer together (on average)? (See p. 316.)		

Refrigerators

Cooling by evaporation is used in a refrigerator to keep food cold. A substance called **freon** is used in **refrigerators**.

1. It is pumped round a circuit of pipes and spends part of its time as a liquid and part as a vapour.

2. As it passes through the pipes outside the refrigerator, it is compressed and turns into a liquid. It warms up a little as it does so and cooling fins help to remove this heat. (Feel the pipes and you will notice they are warm.)

3. The liquid freon then squirts through a valve into pipes that surround the freezing compartment. This is inside the refrigerator.

4. The liquid evaporates, taking the heat it needs from the metal and air inside the cabinet. The vapour is then drawn into the pump and pumped round the circuit again.

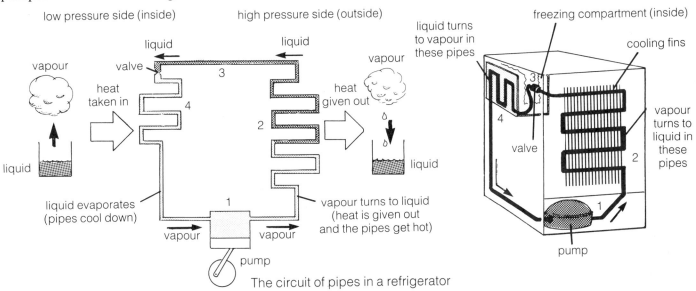

The circuit of pipes in a refrigerator

10 Refrigerators and electric irons are controlled by thermostats. Copy the table below and write in the boxes whether the thermostat is on or off.

	Electric iron thermostat	Refrigerator thermostat
Temperature too high		
Temperature too low		

11 Why is it important to insulate the cabinet of a refrigerator?

12 What happens to the energy that is taken from the inside of a refrigerator?

13 If you leave the door of a refrigerator open, it will **not** cool the room. Why not?

Melting

EXPERIMENT. Melting ice

Melt some crushed ice in a beaker. Use a bunsen with a small flame and take the temperature of the ice every half minute. Stir gently before you take each reading. Plot a temperature/time graph of your readings.

14 The temperature of the ice and water stays around 0 °C for quite a long time. What is the heat energy from the flame doing during this time?

EXPERIMENT. Making very low temperatures

Crush some ice and measure its temperature. Add one teaspoon of salt. Stir the mixture and measure the temperature again. What happens to the ice?

How low can you make the temperature? How does salt affect the freezing point of water?

Explanation. Salt water freezes below 0 °C. Salt sprinkled on ice makes the ice melt. Melting needs heat and so heat is taken from the rest of the salty ice mixture. A salt/ice mixture is a handy way to produce very low temperatures.

15 Why is salt sometimes spread on roads in the winter?

16 Ice sometimes forms on the outside of a beaker of salt and ice. Where does this ice come from?

EXPERIMENT. Melting and freezing wax

Melting. Boil some water in a beaker. Put a tube containing wax into the water and start the clock. Read the temperature of the wax every half minute until it has melted.

Freezing. Take the tube of liquid wax out of the boiling water and take the temperature of the wax every minute as it cools in the air. Plot a temperature/time graph of your readings. Put arrows on the graph to show where the wax is melting and freezing.

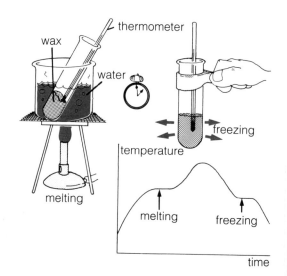

17 Copy and complete

	Melting	Freezing
What changes of state take place?		
Does energy have to be supplied or removed?		

18 The graph below shows how the temperature of a pure crystalline substance changes as it is heated. Copy the graph and label it to show:
(a) when the substance is melting;
(b) when the substance is boiling;
(c) when it is a solid;
(d) when it is a liquid.

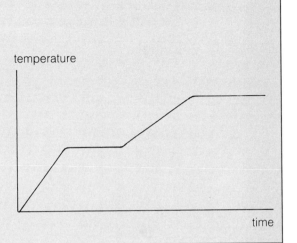

Measuring the energy needed to heat materials

Much of the energy we 'consume' is used for heating, lighting and cooking. So it is useful to know how much energy it takes to heat up materials such as the water in a radiator or the metal in a saucepan. The energy measurements that follow use an electric heater connected to a joulemeter. This measures the amount of energy that flows through to the heater. It is rather like a petrol pump that measures the amount of petrol flowing through. The amount of energy is measured in **joules** (p. 272).

EXPERIMENT. Finding the amount of heat energy needed to warm up a 1 kg block of copper by 10 °C

Put an electric heater into the block of copper. Read the joulemeter and the temperature of the block. Switch the heater on and off until the block has warmed up 10 °C. From the new joulemeter reading find the amount of energy you have used.

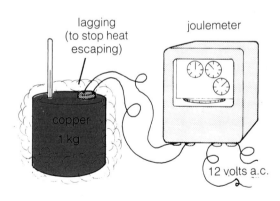

Measuring energy with a joulemeter

19 In an experiment just like this, a student found he had used 3800 joules of energy to produce a 10 °C rise in temperature of the copper block. How much heat energy would be needed to warm the block by only 1 °C?

Specific heat capacity

The **specific heat capacity** of a substance is the amount of energy needed to make 1 kg of that substance 1 °C hotter. The student in the question found that 1 kg of copper needed 380 joules of energy to raise its temperature by 1 °C, so 380 J/kg °C is his value for the specific heat capacity of copper (note the unit used). It is a useful number to know because it tells us how much energy we have to spend to warm up that substance.

EXPERIMENT. Measuring the specific heat capacity of water

Use the method above to find out the amount of heat energy needed to warm 1 kg of water by 10 °C. You can then easily work out the amount of heat energy needed for a 1 °C temperature rise. It is important to keep the heater under the water and to stir well before reading the thermometer. Does it take more heat energy to warm the water than the copper?

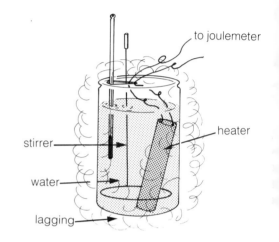

20 Write down the specific heat capacities of these five materials in order, using the proper units.

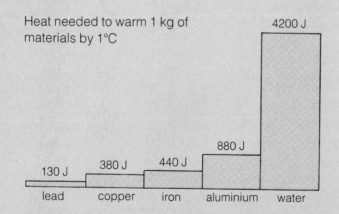

21 Does water need more or less heat per kilogram than copper for the same rise in temperature?

22 Which gives out most heat energy when it cools by 1 °C, 1 kg of copper or 1 kg of water?

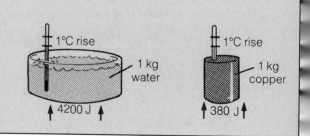

23 Would it take more or less time to boil a kettle of water if the specific heat capacity of water were the same as copper?

24 What properties should a metal have that is going to be used to make saucepans. Copy the table and write in high, low, good or bad.

Density	Specific heat capacity	Type of conductor	Cost

25 A copper saucepan with a mass of 1 kg contains 2 kg of water at 20 °C. How much energy is needed to boil the water, not including energy lost to the surroundings?

The heating equation

If 1 kg of copper needs 380 J for a 1 °C temperature rise, then 2 kg would need twice as much heat for the same temperature rise (2 × 380 J), and if the temperature rise is to be 4 °C this would need four times as much heat energy again (2 × 380 × 4 J). The heat needed can be worked out like this:

heat needed = mass × specific heat capacity × temperature change

J = kg × J/kg °C × °C

Example

How much heat energy must you use to warm 200 kg of water for a bath from 20 °C to 50 °C?
Mass = 200 kg, specific heat capacity = 4200 J/kg °C, change in temperature = 30 °C.
So heat energy needed = 200 × 4200 × 30 = 25 200 000 J.

(a) 3420 J
(b) 4400 J
(c) 5 °C
(d) 125 J/kg°C

26 Copy this table and try and work out the missing numbers.

Material	Mass (kg)	Specific heat capacity (J/kg °C)	Temperature change (°C)	Heat energy needed (J)
Copper	1	380	9	(a)
Iron	½	440	20	(b)
Water	2	4200	(c)	42 000
Lead	1	(d)	10	1250

Measuring latent heat

We found (p. 308) that ice needs latent heat before it will melt. This energy does not make the ice hotter but changes it from a solid into a liquid. Latent heat (a different amount) is also needed to turn boiling water into steam (p. 304).

EXPERIMENT. Finding how much energy is needed to melt 1 kg of ice

Put a heater into a funnel and pack it round with small pieces of ice. Put a beaker and balance under the funnel to collect the melted ice. Switch on the heater and leave it until the balance shows that 100 g of ice has melted. Use the joulemeter to measure the amount of energy you have used to melt this much ice. (You will have to read the joulemeter before and after the experiment.) Calculate the energy that is needed to melt 1 kg of ice.

Example

A heater used 34 000 J of energy to melt 100 g of ice. How much is needed to melt 1 kg of ice?

34 000 J of energy melt 100 g (= 1/10 kg) of ice. Therefore 34 000 × 10 J would be needed to melt 1 kg. So the energy needed = 340 000 J for 1 kg. This quantity is called the **specific latent heat fusion of water.**

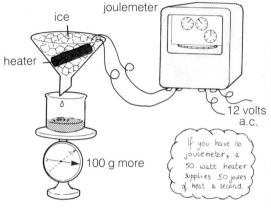

Measuring the latent heat of ice

27 How much energy is needed to melt 5 kg of ice at 0 °C?

28 How much energy must be taken away to freeze 50 g of ice-cold water?
(Specific latent heat of fusion of water = 340 000 J/kg)

29 A thoughtful student was doing the experiment above but used two funnels as shown. He ran the experiment until beaker B had collected 100 g of water more than beaker A. What was the purpose of the second funnel of ice?

EXPERIMENT. Finding how much energy is needed to change 1 kg of boiling water into steam

Use a 3000 W electric kettle to boil some water. Put the kettle on a balance. When the kettle is boiling freely note the balance reading and start the clock. Let the water boil until the balance shows that 100 g of steam have been produced. Stop the clock at this point and note the time taken for this to happen. The energy supplied by the kettle = 3000 × time in seconds (p. 231). This energy changes 100 g of boiling water into steam. From your readings calculate the amount of energy needed to change 1 kg of boiling water into steam.

Measuring the latent heat of steam

Example

A heater used 240 000 J of energy when it changed 100 g of boiling water into steam. How much energy is needed to change 1 kg of boiling water into steam?

It took 240 000 J of energy to boil away 100 g (1/10 kg). Therefore 240 000 × 10 J would be needed to boil away 1 kg.
So the energy needed = 2 400 000 J for 1 kg.
This quantity is called the **specific latent heat of vaporization of water.**

30 How long would it take $\frac{1}{2}$ kg of boiling water, in a saucepan, to boil dry if the heater it is on supplies 1200 J of energy to the water each second?

31 How much energy has to be removed to condense 100 g of steam into boiling water?
(Take the specific latent heat of vaporization of water to be 2 400 000 J/kg.)

32 (a) If you were boiling potatoes on a cooker and the specific latent heat of vaporization of water suddenly dropped to a much smaller value, what two differences would you notice?
(b) Explain one difference such a drop would make to ponds and puddles of water. Do you think it would affect the weather?

33 Explain why you would have to eat your lollipop faster if the specific latent heat of fusion of water was much smaller than it really is.

34 The kinetic theory of matter

The particles of matter

What are the materials made of and what happens to them when they change state? To find out we must examine them in different ways. An electron microscope can magnify 80 000 times, and materials looked at in such detail seem to be made up of small particles. A beach looks smooth from a distance but is seen to be made of particles when looked at close up.

Electron micrograph showing protein molecules

A model of a gas

Some well-known facts about gases are:
1. A gas can be compressed.
2. A gas is 'thin'. It has a low density.
3. When a substance turns into a gas, there is a great increase in its volume.

These observations can be explained if we imagine a gas is made of particles that are spaced out far from each other with nothing between.

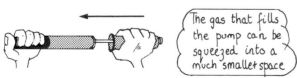
A gas can be compressed

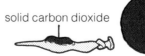
The particles of gas are spaced out

EXPERIMENT. To show that the particles of a gas are moving

a) Place a dish of after-shave, scent or ammonia solution on the bench. The molecules will soon reach your nose.

b) When hydrogen chloride gas meets ammonia gas, a white smoke is formed. Place a drop of concentrated hydrochloric acid at one end of the tube and a drop of ammonia solution at the other. After a while white smoke appears some distance from the ends of the tube.

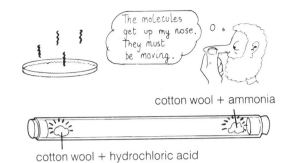

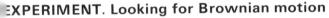

1 What do these experiments suggest about the molecules of a gas?

2 Solids and liquids cannot be compressed like a gas and have a much greater density. What does this suggest about the particles of solids and liquids?

EXPERIMENT. Looking for Brownian motion

The movement of gas particles can be shown up by looking at smoke through a microscope. The smoke has to be held in a small plastic cell and lit from the side by a lamp. Smoke is made of tiny pieces of soot floating about in warm air. Even under the microscope these pieces of soot only look like golden points of light. Watch the specks of soot carefully and you should see them move. How would you describe this movement, that was first reported by Robert Brown in 1827.

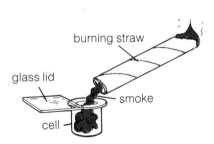

Filling a smoke cell

3 How can you be sure that the small movements of the soot specks are not caused by the light?
Could the movement be caused by the smoke swirling about in the cell?

4 The moving specks of light are:
  A. Air particles moving in a regular way.
  B. Air particles moving in a haphazard way.
  C. Specks of soot colliding.
  D. Specks of soot moving in a regular way.
  E. Specks of soot moving in a haphazard way.
  Which is the most likely?

What causes the motion?

The specks of soot are suspended in the air and are seen to move continuously in small haphazard jumps. This movement is thought to be caused by the millions of air particles that surround the speck. They move and strike it from all directions but not in equal numbers. The unequal bombardment then causes the speck of soot to move first one way then the other.

This movement of the specks of soot is evidence that the invisible air molecules around them are also moving.

speck of soot
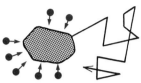

> **5** What does this experiment tell us about the particles of a gas?
>
> **6** At higher temperatures, the specks of soot move further and faster. What else does this tell us about the movement of the gas particles?
>
> **7** Small bits of carbon, floating in a liquid, show Brownian motion. What does this tell us about the particles of a liquid?

EXPERIMENT. To show that the particles of a liquid are moving

Carefully pour blue copper sulphate solution into water in a beaker, so that it forms a separate layer at the bottom. Use a funnel to add the blue liquid and build up the layer gradually from underneath.

Leave the beaker to stand and watch the boundary between the two liquids. After a while you should see the clear boundary become blurred as blue liquid starts to move up and mix with the water.

The movement of one liquid into another like this is called **diffusion** and is evidence that the particles of a liquid move about.

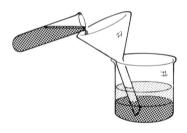

Heat and kinetic theory

The idea that the particles of a substance are moving is called the **kinetic theory**. The particles of a solid are pictured as vibrating rapidly but keeping their positions in neat rows, held together by strong forces of attraction. According to this theory, even the particles of a cold solid are vibrating. To stop the vibrations completely, the solid would have to be cooled to 273 °C below zero. There is nowhere on Earth as cold as this.

When a solid is heated the vibrations increase until they are so violent that the forces can no longer hold the particles together. The neat rows of particles break up and the solid melts, forming a liquid.

When a liquid is boiled the energy of the particles becomes so great that they are able to fly off into the space above the liquid, forming a gas. A gas is pictured as countless millions of tiny particles moving in space at speeds of about 1000 km/hour, bouncing off each other and anything else in their way.

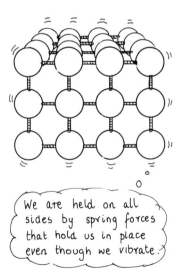

A model of the particles of a solid

8 Copy and put solid, liquid or gas in the spaces.

What the particles are doing	The state
The particles are free from each other and moving at high speeds.	
The particles are vibrating and are loosely held together but not in a neat pattern.	
The particles are vibrating and held in neat rows.	

9 As steam changes to water and then to ice, does the energy of its particles increase or decrease?

Solid to liquid to gas – a model of what happens to a substance when it changes state

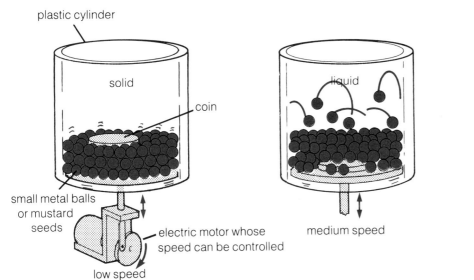
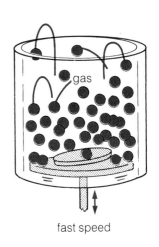

A model of the changes that take place when a substance changes state

10 Describe the changes that take place as the balls are given more and more energy.

11 What would you feel if you put your hand into the cylinder?

12 Why do high speed gas particles not break windows and glass bottles?

13 Can we feel the collisions of the particles of gas in an air bed? What would happen if they suddenly stopped moving?

Gas pressure and the kinetic theory

We have seen that a gas exerts a pressure that increases if the gas is heated or compressed. The kinetic theory can explain why this happens.

Questions	Answers
1. Why does a gas exert a pressure?	1. The particles of a gas move quickly and there are millions of particles present. So there is an enormous number of tiny collisions between the particles and the container each second. This constant bombardment of the walls is why a gas exerts a pressure on the walls.
2. Why does the pressure go up if the volume goes down?	2. If the gas is compressed, the particles have less volume to move in. They collide more often with the walls and the pressure of the gas rises.
3. Why does the pressure go up if the temperature goes up?	3. If the gas is heated, the particles move more quickly. So they collide more often with the walls and the pressure of the gas increases.

14 Copy the table and insert 'no change' or 'increases' into the boxes.

	Reducing the volume of the gas	Increasing the temperature of a gas
What happens to the speed of the particles?		
What happens to the number of collisions per second?		
What happens to the pressure of the gas?		

If the gas is cooled down, the particles slow down. At the absolute zero of temperature (p. 165) the molecules would have lost all their energy and the pressure of the gas would be zero.

How gases behave – a summary of the gas laws

If a gas is compressed or heated, its pressure, volume and temperature usually all alter. For most gases the changes in pressure, volume and temperature follow simple laws. To look for the laws experimentally each of the three quantities must be kept constant while the other two are altered (see pp. 164, 165, 157).

This leads to three gas laws that are summarized here.

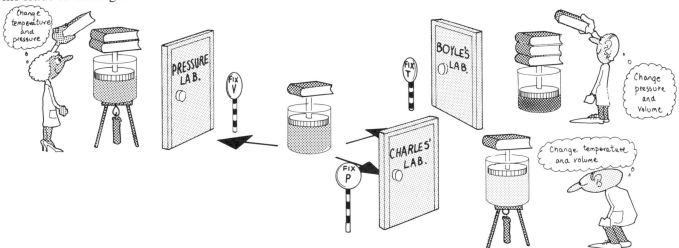

Charles' law (*P* fixed, *V* varies with *T*)

The volume of a fixed mass of gas is proportional to its Kelvin temperature, provided the pressure is constant.

15 Do these results follow Charles' law?

Volume, V (mm³)	70	75	80	85	90
Temperature, t (°C)	7	27	47	67	87

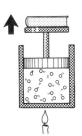

AS THE TEMPERATURE RISES THE MOLECULES GAIN ENERGY. THEY MOVE FASTER AND COLLIDE MORE OFTEN. THE PISTON MOVES UP AND THE GAS EXPANDS.

Boyle's law (*T* fixed, *V* varies with *P*)

The pressure of a fixed mass of gas × its volume is a constant provided its temperature does not change.

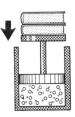

WHEN THE VOLUME OF A GAS GETS SMALLER, ITS MOLECULES (TRAVELLING AT THE SAME AVERAGE SPEED) HIT THE WALLS MORE OFTEN. THIS INCREASES THE PRESSURE OF THE GAS.

The Pressure law (*V* fixed, *P* varies with *T*)

The pressure of a fixed mass of gas is proportional to its Kelvin temperature provided its volume does not change.

16 Do these results follow the Pressure law?

Pressure, P (kPa)	67.5	75	82.5	90	93.3
Temperature, t (°C)	−3	27	57	87	100

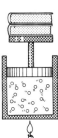

AS THE TEMPERATURE RISES THE MOLECULES GAIN ENERGY AND MOVE FASTER. THE VOLUME IS FIXED SO THE MOLECULES COLLIDE WITH THE WALLS MORE OFTEN AND THE PRESSURE RISES

35 Radioactivity and the atom

A model of the atom

Atoms are far too small to see and so we invent models or mental pictures of what they are like. These models have to fit in with the findings of experiments and from time to time have to be altered to take account of new discoveries. One useful picture of the atom is the Bohr model.

This sees every atom as being built from three basic particles: electrons, protons and neutrons. The protons and neutrons are bound together by strong nuclear force to form a tiny nucleus at the centre of the atom. This nucleus has nearly all of the mass of the atom and the protons give it a positive charge. The electrons move in orbits further out and surround the nucleus with a light shell of negative charge. In a neutral atom, the charges on the nucleus and the electrons cancel out.

Building the elements

All the atoms found in nature have between 1 and 92 protons in their nucleus. All except hydrogen also have a number of neutrons that help hold the nucleus together against the repulsive force of the protons. The number of protons in a nucleus is called its proton number (Z) and the total number of particles (protons+neutrons) is called nucleon number (A). The nucleus will have Z units of + charge from its protons and a neutral atom will have Z electrons in orbit outside the nucleus to cancel this charge.

An element is a substance made up of atoms that all have the same number of protons and so the proton number of an atom determines what element it is. All this information about a nucleus can be written as the symbol $^A_Z E$ where E is the chemical symbol of the element.

(Atoms of an element that all have the same number of protons but different numbers of neutrons in their nuclei, are called **isotopes** of the element.)

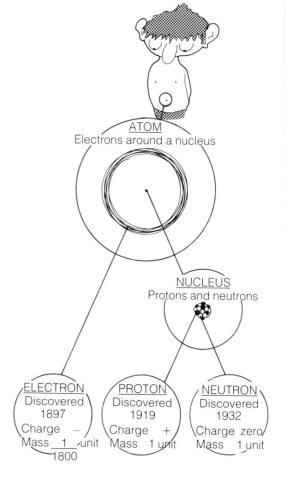

Nucleus	Element	Proton number	Nucleon number	Nucleus	Element	Proton number	Nucleon number
+ ●	Hydrogen ($^1_1 H$)	1	1	+++ ●●●	Lithium ($^7_3 Li$)	3	7
++ ●●	Helium ($^4_2 He$)	2	4	++++ ●●●●	Beryllium ($^9_4 Be$)	4	9

● proton
○ neutron

The nuclei of the first four elements

1. Copy this table of definitions and write the correct terms in the blank spaces. The terms are proton number, nucleon number, element, nucleus, atom, proton, neutron, electron, isotope.

Term	Definition
	A substance made from only one type of atom.
	The smallest piece of an element.
	The central part of an atom containing protons and neutrons.
	A subatomic particle that carries a + charge.
	A subatomic particle that carries no charge.
	Number of protons in a nucleus.
	Number of particles in a nucleus.
	A particle with one unit of − charge and much less mass than a proton.
	Atoms with the same number of protons but different numbers of neutrons in their nuclei.

2. Why must there be a strong attractive force between the nuclear particles? What would happen to a nucleus if there was no nuclear force?

Radioactivity – radiation from the atomic nucleus

Radioactivity was discovered in 1896 by a scientist named Becquerel. He was looking for X-rays from uranium and found a new type of radiation. The uranium atoms were radioactive. He found that rays from uranium could expose photographic plates, even in the dark. Uranium was the first element that was found to be radioactive but a new more active element (radium) was discovered soon after by the Curies.

DEMONSTRATION. Detecting radioactivity with a Geiger–Muller tube

A convenient way of detecting radioactivity is with a Geiger–Müller tube and scaler. (The scaler is an instrument that usually contains an amplifier, a voltage supply for the G-M tube and a counter.)

Connect the G-M tube to the scaler and point it at a radioactive source. Turn up the voltage to the tube until the scaler begins to count and click. Notice how the equipment shows up the invisible and silent radiation coming continuously from the element.

Measure the number of radiation counts that are collected by the tube in 1 minute. Then without moving the element or tube measure the count again. Repeat this a number of times to see if you get the same count each minute.

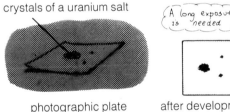

photographic plate (in the dark)   after development

Radioactive radiation can expose photographic paper

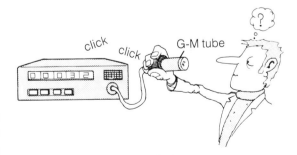

Detecting radiation

3 Radioactivity is rather like sparks and fireballs shooting out of a firework in fits and starts. Name one difference between getting a firework and getting a radioactive element to give out its radiation. Name two other differences between the radiations.

4 These numbers show the counts/minute measured by a G-M tube from a radioactive element.
   (a) What is the average count/minute?
   (b) What is the greatest difference from that average?
   (c) Why do you think that the numbers are different? Is it because
      (i) of experimental errors of measurement,
      (ii) different numbers of particles leaving the element each minute.

| Counts/minute | 3630 | 3570 | 3540 | 3660 | 3550 | 3650 | 3610 | 3590 |

How the Geiger–Müller tube works

The tube contains a wire at a high voltage (about 450 V) inside a metal case that is earthed. The voltage of the wire is made just too low to cause a spark between the wire and the case. Then, when radiation enters the tube, it triggers off a quiet spark. An amplifier and loudspeaker connected to the tube click every time there is a spark, showing the presence of radioactivity.

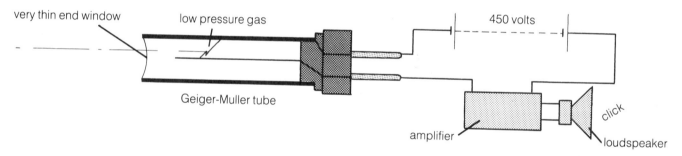

DEMONSTRATION EXPERIMENT. Radioactivity is all around us

Remove the radioactive source of the previous experiment to another room. The scaler will continue to count slowly as radiation is picked up by the tube. Move the tube around (carefully) to try to find out where this radiation comes from. Is there anywhere free from the radiation?

Background count

Use a clock with the apparatus to measure the number of counts this 'background radiation' gives in one minute. Is it always the same number each minute?

5 A student took readings of the background count for times of one minute. Here are her results:

| Counts/minute | 27 | 19 | 25 | 28 | 15 | 23 | 16 | 20 | 26 | 11 |

She had not altered her apparatus in any way, but her results were all different. Did she make a mistake or can you explain why her results are like this? Find the average background count rate.

6 Our world is radioactive and has been for a very long time. Some rocks on the Earth are radioactive, radioactive rays shower down on us from space, and man has added a little to nature's radioactivity. Which of these statements contains the most truth:
   (a) Background radioactivity will kill us.
   (b) Background radioactivity does cause injury but so few people are affected that we accept it as normal.
   (c) Radioactivity is completely harmless to our bodies (p. 329).

Is there more than one sort of radioactive radiation?

The early atomic scientists found that three different sorts of radiation came from radioactive materials. Each had a different penetrating power. They called the radiation alpha (α), beta (β) and gamma (γ) rays.

DEMONSTRATION EXPERIMENT. Finding what materials are needed to stop alpha, beta and gamma rays

Line up an americium source with a G-M tube and scaler. (Americium gives out alpha rays.) Place a sheet of paper between the source and the tube to see if alpha rays can pass through paper. The scaler counts much slower, showing that very little radiation can get through. Replace the americium with strontium-90 (this gives out beta rays) and then cobalt-60 (this gives out gamma rays). Use thicker and thicker plates of aluminium and then lead to find what it takes to stop each radiation. You should find the following:

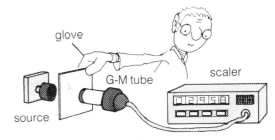

Testing for alpha, beta or gamma rays

Type of radiation	Material needed to stop it
Alpha	A sheet of paper (or 3 cm of air)
Beta	3 mm of aluminium
Gamma	3 cm of lead

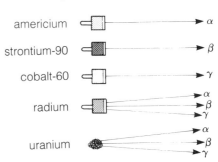

Note that some radioactive materials give out all three types of radiation.

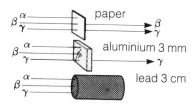

What is needed to stop each type of radiation

7 Which radiation:
(a) gets rid of its energy quickest?
(b) could not get through skin very easily?
(c) could go furthest through the air?

8 Copy the table, put a ✓ in the square if the radiation can get through; an X if it cannot.

Type	Paper	3 mm aluminium	3 cm lead
Alpha			
Beta			
Gamma			

9 Radiation from some elements contains more than one type. Work out which types of radiation are present from the following readings (all corrected for background count).

Element	Count/minute with nothing between source and counter	Count with paper in the way	Count with 3 mm of aluminium in the way	Count with 3 cm lead in the way	Types of radiation present
A	3500	2000	2000	0	
B	3500	3500	2000	0	
C	3500	2000	20	0	

What are the radiations made of?

Careful experiments by the early atomic scientists found that the radioactive radiations had the following properties:

Type	Effect of a magnetic field	What the radiation is made from
Alpha	Deflected like a 'heavy' positive particle	The nucleus of a helium atom (i.e. 2p+2n)
Beta	Deflected like an electron	A high-speed electron
Gamma	Not deflected at all	Electromagnetic radiation with even more energy than X-rays

Where do the radiations come from?

Experimental work by Rutherford and Soddy in the early 1900's helped them to discover the origin of these radiations. They showed that the radiation came from inside the atom; from its tiny central **nucleus**. The nucleus of a radioactive atom is unstable and can suddenly break up, throwing out high-speed particles and radiation. The radiation we detect is a result of these nuclear events.

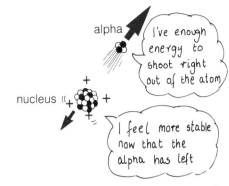

An alpha particle leaves the nucleus of a decaying atom

Ionization by radiation

When radioactive particles streak through air they are able to remove electrons from some of its atoms. Alpha particles do this best because they are much heavier than the electrons and carry a positive charge. Beta particles and gamma ray photons are lighter than alpha particles and so are not so good at removing electrons from atoms.

Atoms that have lost electrons have a net positive charge and are called positive ions. Radiation leaves trails of ions and electrons through the air making it a good conductor of electricity.

When radiation enters our bodies it ionizes atoms and causes chemical changes. Alpha particles cause most damage because they produce large numbers of ions within a short distance. Beta and gamma rays penetrate further because they produce fewer ions along their tracks and take longer to use up their energy.

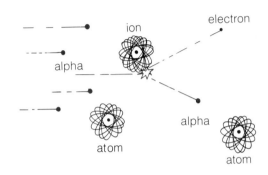

Alpha particles ionizing atoms

Chance happenings

EXPERIMENT

Warm a little cooking oil in a frying pan and drop in a seed of 'popping corn'. Take cover and try to guess when the corn is going to explode. Throw in a handful of corn and watch. Do all the seeds explode together and can you tell which one will explode first? We know that all the seeds will explode eventually but we cannot tell when a particular seed is going to change.

> **10** Explain why the 'activity' of the corn is fast at first but gets less as time goes on.

Uranium atoms are unstable and we know they will break up sooner or later. But they do not all break up (decay) at once. We cannot tell which atoms are about to decay, nor can we force an atom to decay. Radioactive decay is a game of chance that nature plays, completely outside of our control. When uranium decays it changes into a different substance with different properties. So the number of uranium atoms in the world gets less and less as time goes by.

Half-life

In a pan full of popping-corn, there are a large number of seeds that can explode and the 'activity' is high. This means that the number of unexploded seeds drops rapidly. After a time, when there are fewer seeds left to pop, the activity becomes less and the few seeds that remain die away more slowly.

A sample of radioactive atoms behaves in a similar way except there are many millions more of them. At first the radioactivity of the sample is high and the number of active atoms drops rapidly. Later, with fewer atoms left to decay, the activity drops and the active atoms that remain, die away more slowly. The lifetime of the sample is almost infinite and so the idea of **half-life** is used instead. Half-life is the time it takes for half of the number of atoms in a sample to decay. (Note that for half of a sample to decay to one quarter also takes one half-life.)

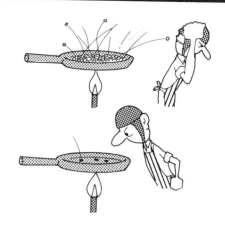

EXPERIMENT The decay of 'radioactive' dice

Take 50 small wooden cubes (or dice) and paint one face of each black. The cubes represent radioactive atoms and those that show a black face when thrown are supposed to have 'decayed'. Throw all the dice together and take out those that, by chance, have decayed. Count how many dice are left after one throw. Throw these again and take out the cubes that have decayed this time. Carry on throwing and counting until all of the dice have decayed. You will notice how much slower the cubes decay as their number decreases. Plot a graph of the *number of dice left* against *the number of throws*, to give a picture of the decay of the dice.

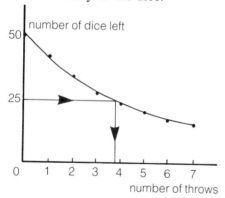

Finding the 'half-life' of dice

---

11 Plot a decay curve of these results from a 'radioactive' dice experiment and from it find the number of shakes it takes to reduce the number of dice by half (their half-life).

Number of dice left	50	42	34	27	23	20	17	15
Number of shakes	0	1	2	3	4	5	6	7

12 If you repeated the experiment why is it unlikely that you would get the same results?

13 How could you make your dice decay more quickly, i.e. make them more radioactive.

EXPERIMENT. The half-life of radon (radon is a radioactive gas)

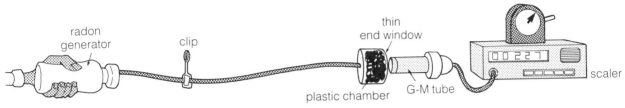

Radon has a short half-life and has to be generated by the decay of thoron in a plastic bottle. Squeeze some radon gas from its generator bottle into a container with a thin front window. Clip the connecting tube, to isolate the gas, and fix a G-M tube close to the window. Start the scaler and clock; stop and read the scaler after 10 s but leave the clock running. When the clock reaches 30 s, switch on the scaler again and find the count at the end of the next 10 s. Carry on counting for 10 s every half minute as the gas decays. Plot a graph of *counts/10 s* against *time* and from it find the half-life of radon.

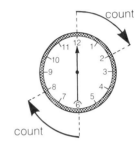

14 Here are results from such an experiment.

Time (minutes)	0	$\frac{1}{2}$	1	$1\frac{1}{2}$	2	$2\frac{1}{2}$	3
Counts in 10 s	200	138	100	71	50	35	25

Plot a graph and from it find the half-life of radon. With this experiment the count never goes down to zero. Why not?

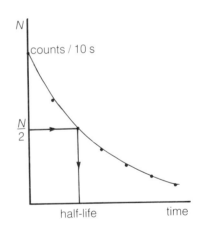

Other half-lives
Carbon-14 5730 years, uranium-238 4500 million years, plutonium 25 000 years, radon-220 56 seconds, lead-214 27 minutes.

15 (a) Give one reason why the disposal of radioactive materials like plutonium is such a problem.
(b) Why wouldn't a lead-214 salesman be a very good job to have?
(c) Why can't we buy radon ready made from a supplier? (We have to generate our own when we want it.)

Heat from radioactivity – nuclear fission

A special form of uranium can be used in nuclear power stations to produce heat and turn water into steam. The heat is produced when the atoms of uranium split in two – a process that is called nuclear fission. Normally uranium atoms decay by sending out alpha and beta particles. These are small compared with the uranium nucleus. A different thing happens if a **neutron** enters the nucleus. The nucleus breaks up into two nearly equal parts and three more neutrons. Energy is released when this happens.

Radioactive decay

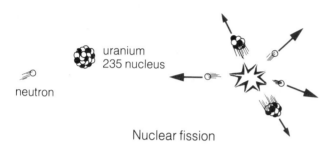

Nuclear fission

Atomic bomb

If these neutrons enter three other nuclei, these would split and produce nine fresh neutrons to continue their work. A run-away 'chain reaction' would set in and the uranium would explode. This process is what happens in an atomic bomb.

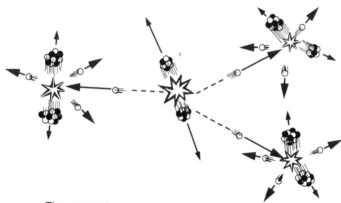

The start of an uncontrolled fission chain reaction

Nuclear power

In nuclear power stations two of the three neutrons released are carefully absorbed, leaving one to carry on and split up the next uranium nucleus. The fission process then ticks over steadily, producing heat in a controlled way. A few kilograms of uranium can produce as much heat as hundreds of tonnes of coal in this way. The uranium and other fuels used in nuclear power stations are, however, very dangerous materials. They can escape and spread radioactive substances into the atmosphere. They also produce radioactive waste that is difficult to dispose of.

What radiation can do to our bodies

Very intense radiation...
...as from the explosion of nuclear weapons or accidents in nuclear power stations.

What happens to people who are exposed to very intense radiation was seen soon after atomic bombs were exploded over Hiroshima and Nagasaki in 1945. Many died immediately from the shock wave and intense heat of the blast. Others survived the blast but received massive doses of radiation that gave them radiation sickness. This is a high fever with vomiting and diarrhoea that dries out the body and brings death within a few days.

People who had lower doses survived the fever but found they had no resistance to normal disease. The radiation damaged bone marrow cells and destroyed the healing properties of the blood. Wounds and burns would not heal and mild diseases like tonsilitis caused death. These injuries and deaths occurred soon after the explosions.

The radiation also caused damage to the victims that has had a long-term effect and is still causing cancers 30 or 40 years after the event.

Nuclear explosions produce a 'mushroom' cloud

Radiation hits us all
Radiation strikes our bodies even without nuclear explosions. Background radiation (p. 322) comes from space, rocks in the Earth and from man's use of radioactive materials. Some examples of these are shown in the chart.

Fallout	Radioactive waste	Medical uses	Uses in factories	Use at home
From nuclear explosions and emissions from nuclear and coal-fired power stations.	From nuclear reactors.	Treatment of disease and research.	On production lines to check if packets are full or if thickness changes or the level of liquid or powder in a container. Sterilization of food.	Luminous watches, gun sights, exit signs, old false teeth, trimphones, camping gas mantles, smoke detectors, starters for tube lights, smoke from coal fires...

The energy of the radioactive particles that make up background radiation is absorbed by the cells of the body. These can be damaged and cause diseases such as leukaemia (cancer of the blood), cataract (causing blindness), cancers and sterility. Alpha particles and neutrons cause about 10 times as much damage as beta or gamma rays.

Background radiation delivers an average dose of 1.9 mJ of energy to each kilogram of our bodies every year and about 2000 people a year will get cancers from its effects.

Radiation strikes from inside

Radioactive materials are taken into our bodies as we eat and breathe. (The picture shows examples of how iodine-131 (from 'fall-out') and radon gas (from rocks) gets into our bodies.) Some of these materials stay and collect in the bones and organs, gradually getting more concentrated. The radioactive organs then radiate us from inside our bodies.

Fast-growing cells

Cells that are multiplying quickly are more likely to be damaged by radiation than normal cells. A fast-growing baby in the womb of its mother is especially at risk. Radiation and X-rays could kill the baby or deform it in some way.

Cancer cells also grow rapidly and radiation is widely used to kill cancerous growths. Doses of radiation (usually gamma rays) are aimed at the growth until it is destroyed.

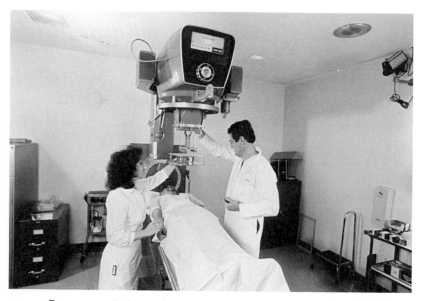

Gamma radiation unit used to destroy cancerous growths

Genetic effects

Radiation can kill body cells but it can also have a more sinister effect. It can alter the nucleus of cells. The nucleus contains the cell's instructions for making new life. Radiation can interfere with these instructions so that children are born with mutations or handicaps. Radioactivity is not the only cause of genetic defects, but the more radioactivity there is around, the more common such abnormalities will become.

All radioactive radiation is dangerous and there is no safe level. Even gentle doses of radiation spark off changes, that many years later, cause disease or handicap. As we step up the use of radioactive materials, more people will become ill or die from the effects of the radiation. This fact has to be balanced against the good that radioactivity can do.

Safety

Radioactive materials must be handled with great care. You should not touch them or point them at the body, especially the eyes and the sex organs. They should be held in tweezers at arm's length and stored in a lead-lined box, in an out-of-the-way place. Do not take any risk with this invisible radiation. Damage it causes may not show up for several years, or until you have children.

---

16 Make a list of ten of the effects of radiation mentioned above.

17 Rewrite your list under two headings: long-term effects; short-term effects.

18 Put a * by any good effects.

The units of radiation dose

The energy absorbed by matter when radiation falls on it is measured in grays. A dose of 1 gray is 1 joule of energy per kilogram.

Equal doses of different types of radiation do not in general have the same biological effect. Alpha particles and neutrons cause more damage than gamma rays or beta particles. Each type of radiation is given a **quality factor** to allow for the amount of damage it does.

Gamma rays	1
X-rays	1
Beta particles	1
Neutrons	10
Alpha particles	20

(these figures depend on particle speed)

Quality factor multiplied by the dose gives the **dose equivalent**. This is measured in sievert (Sv) or millisievert (mSv)

Some organs in the body are more easily damaged than others and the same dose of radiation causes different risks to different organs. To allow for this the tissues and organs are given a risk factor. The dose (equivalent) to each organ is multiplied by the risk factor and added to give the dose equivalent for the whole body. This is called the **effective dose equivalent** and is measured in sieverts.

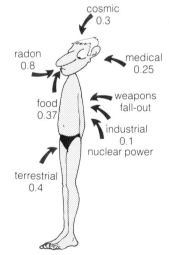

One year's dose of background radiation in mSv

Some special dose levels

The amount of radiation we receive varies from place to place. Here are some examples of doses of radiation that were received by people in special situations.

The world record for a home contaminated by radon, Philadelphia, USA – 400 mSv per year.
The highest dose from radon in a house in the UK – 320 mSv per year.
The highest dose to a child's thyroid gland from fallout after a fire in the Windscale nuclear power station (1957) – 160 mSv per year.
Maximum recommended dose to the public from the nuclear industries – 1 mSv per year.
Maximum dose allowed for a nuclear worker – 50 mSv per year.
Average dose to a nuclear worker – 4 mSv per year.
Typical dose from a medical diagnosis using technectium-99 as a tracer – 5 mSv.
Average dose from background radiation – 2 mSv per year.
* Maximum dose allowed to the members of the public – 5 mSv per year.
* Note that there is no safe dose level and this limit is chosen so that the risk of injury from radioactive radiation is equal to the risk of an industrial accident.

ASSIGNMENT

Make a bar chart from these figures to show up the range of dose levels that are possible from everyday radioactive radiation.

Measuring activity

The unit used to measure the activity of a material is bequerel (named after the discoverer of radioactivity). 1 bequerel (Bq) is the decay of one nucleus per second.

Radioactive sources used in school laboratories have an activity of about 100 kBq, showing that 100 000 nuclei are decaying each second.

Other examples of activities are:

- Legal limit (Sweden) of caesium-137 in reindeer meat intended for human consumption = 300 Bq/kg.
- Average level (Oct. 86) in reindeer meat after Chernobyl = 8000 Bq/kg.
- Action level for milk in UK = 2000 Bq/litre.
- Action level for milk in Germany = 500 Bq/litre.
- Average level of plutonium in the lungs of dead people in UK = 2 mBq/kg.
- Some levels in the lungs of dead nuclear workers (Sellafield) = from 120 to 1140 mBq/kg.
- Estimated release of iodine-131 from Chernobyl accident = 10^{18} Bq and Caesium = 10^{17} Bq.
- Concentration of caesium-137 at some 'hot spots' (Sweden) from Chernobyl fallout = 200 Bq.
- Average concentration of radon in American homes = 50 mBq/litre.
- Average amount of naturally occurring potassium-40 in an adult = 4000 Bq.

ASSIGNMENT
Rewrite these examples in order of size.

Fallout in the food chain

A story of nuclear pollution.

On 25 April 1986 there was a nuclear accident in a reactor at Chernobyl in Russia. Some radioactive fuel rods exploded when the fission chain reaction ran out of control. The rise in power that resulted together with white hot fragments of the core, produced vast quantities of steam that blew the lid off the reactor and spread radioactive debris around the area. This set off fires and created a very high radiation field around the plant. The core, now open to the air, was left burning and further chemical explosions occurred as water reacted with zirconium pressure tubes to make hydrogen and carbon monoxide.

Nuclear reactors contain many 'fission products' that come from the nuclear reactions. With the shielding gone, these began to get into the atmosphere. First were inert radioactive gases such as xenon-133. They were swept high into the atmosphere by the heat of the fires and blown north by the wind towards Finland and Sweden. Inert gases do not dissolve or react and so cannot enter the food chain. However, anyone out in the open as the cloud passed overhead would be irradiated by it.

Other fission products that got into the atmosphere were iodine-131 and caesium-137. These are volatile elements that were released in the form of fine dust particles or vapour. They were swept up as a radioactive plume into the air and within five days had been spread by winds over most of Europe, Scandinavia and Britain. (Not many of the heavier, long-lived, radioactive elements like plutonium, curium and neptunium fell over Western Europe.)

Chernobyl nuclear power station

The iodine and caesium were washed out of the clouds by rain that fell during those five days, producing radioactive hot spots on the ground. In Britain the highest concentrations of radioactivity were found on high lands such as Wales, Cumbria and Scotland.

The radiation enters the food chain when it is taken up by vegetation that is then eaten by sheep or cows. It appears in milk within a few days and in the meat of the animals after some months. Iodine-131 has a relatively short half-life of eight days but still has time to enter our bodies through radioactive milk. Caesium has a half-life of 30 years and is much more persistent. It can contaminate vegetation and animals for many years at great cost and waste.

The Lapp reindeer herders of Northern Sweden are an example of a group who have been particularly hard hit by the fallout. Rain fell over parts of their homeland as the cloud passed and deposited caesium-137 on the lichen that the reindeer eat. Lichens have no roots and take all their nourishment from the air. They absorb up to three times as much radiation as other plants and having no roots, cannot pass it on to the ground. Many thousand reindeer became radioactive from eating the lichen and their meat had to be condemned. The Lapps traditional life-style and livelihood have been disrupted and will not return to normal until the contaminated lichen has been eaten or has lost its radioactivity. This could take a generation.

The chemical explosion at Chernobyl devastated the plant and the area around but it will do no further damage; the effects of the nuclear explosion were global and will remain a danger to health for years to come.

ASSIGNMENT

Draw a labelled picture that shows how radioactivity from Chernobyl got into reindeer in Sweden and people in Britain.

Matter check list

After studying the chapters on the nature of Matter and Radioactivity you should know that:

- the mass of a body is a measure of the amount of material it contains or its inertia
- volume is a measure of the space taken up by a body
- density is mass/unit volume measured in kg/m^3
- air has a density of about 1.2 kg/m^3 and water 1000 kg/m^3
- the three chief states of matter are solid, liquid and gas
- latent heat is needed to change solid to liquid and liquid to gas
- salt lowers the melting point of ice and that a salt/ice mixture produces low temperatures
- specific heat capacity = the energy needed/kg to warm a material by $1\ °C$
- the heating equation is heat lost (gained) = mass × specific heat capacity × temperature drop (rise)
- specific latent heat is the energy needed/kg to change the state of a material
- evaporation of liquid to gas occurs at any temperature and increases with surface area and draught
- evaporation cools a surface because it draws latent heat from it
- condensation is vapour turning to liquid
- a refrigerator is a heat pump that moves heat from inside an insulated box to the outside
- all matter is made of particles
- the kinetic theory states that the particles are in perpetual motion
- Brownian motion is the random motion of specks of soot under molecular bombardment
- diffusion is the gradual movement of one material through another caused by the movement of their molecules
- the pressure of a gas is due to the continuous bombardment of fast moving molecules
- pressure increases with temperature because the molecules move at a higher average speed
- pressure increases with reduced volume because the molecules strike the walls more frequently
- molecules in a solid vibrate but are held in place by their neighbours
- molecules in a liquid move about and are not fixed in position, but are still attracted by their neighbours
- molecules in a gas are free from each other and move at about the speed of sound
- the Bohr picture of the atom is a small positive nucleus of protons and neutrons surrounded by electrons in set orbits
- the proton number is the number of protons in the nucleus

- the nucleon number is the number of protons and neutrons in the nucleus
- the atoms of an element all have the same proton number
- an isotope of an atom has the same number of protons but different numbers of neutrons
- radioactive emissions come from the nucleus of unstable atoms
- there are three main types of radioactive radiation called alpha (helium nuclei), beta (electrons) and gamma (electromagnetic photons)
- radiation can be detected by a G-M tube and scaler or ratemeter
- paper will stop alpha, 3 mm of aluminium will stop beta and 3 cm of lead will stop gamma rays
- radiation causes the ionization of atoms by stripping off electrons
- alpha particles have about 10 times the ionizing power of beta particles or gamma photons
- radioactive decay is random and gets less as the number of atoms drops
- half-life is the time taken for half a given sample of atoms to decay
- all radioactive radiation is dangerous and can cause damage to the body
- intense radiation can cause sickness and death and less intense radiation can cause long-term damage

Revision quiz

Use these questions to help you revise the work on Radioactivity and the nature of Matter.

- What is the difference between mass and weight? — ...mass is a measure of the amount of material in a body (in kg); weight is the force of gravity on a body (in N).

- Define density. — ...density=mass per cubic metre of a material.

- What must be provided to change the state of a substance? — ...latent heat is needed to change a solid to liquid and liquid to gas.

- Why does mixing salt with ice produce such low temperatures? — ...the salt lowers the melting point of the ice which melts and draws latent heat from the surroundings, cooling them down.

- Define specific heat capacity. — ...the heat energy needed/kilogram to raise the temperature of a material by 1 °C.

- What is the heating equation? — ...heat lost(gained)=mass×specific heat capacity×temperature drop (rise).

- Define specific latent heat. — ...the energy needed/kilogram to change the state of a substance at its melting or boiling point.

- How does the evaporation of perspiration cool the body? — ...when sweat turns to vapour it takes the latent heat it needs from the skin.

- What is the kinetic theory of matter? — ...that the particles of matter are in perpetual motion.

- What experimental evidence is there for the kinetic theory? — ...Brownian motion and diffusion.

- What is the kinetic theory picture of a solid? — ...particles strongly attracted together and vibrating energetically about fixed positions.

- ...and a liquid? — ...particles moving freely and yet still held by attraction for each other.

- ...and a gas? — ...particles free from each other, moving and colliding at high speeds.

- Why does a gas exert a pressure on a surface? — ...gas pressure is due to the continuous bombardment of the surface by millions of high-speed molecules.

- Why does gas pressure increase with temperature? — ...because the molecules move at a faster average speed and strike the walls with more force.

- Why does gas pressure increase as volume is reduced? — ...because the molecules strike the walls more frequently.

- How does the kinetic theory explain latent heat? — ...gas molecules have more energy on average than liquid molecules and so energy has to be supplied to change the liquid to gas.

- What is the Bohr model of the atom? — ...a small, very dense nucleus, with a + charge, made of protons and neutrons and surrounded by electrons in set orbits.

- What is the proton number of an element? — ...the number of protons in the nuclei of its atoms.

- What is the nucleon number of an element? — ...the number of protons+neutrons (nucleons) in the nuclei of its atoms.

- What is an isotope of an element? — ...it has atoms with the same number of protons in their nuclei but different number of neutrons.

- Where do alpha, beta and gamma rays come from?

- What are alpha rays?

- What will stop most alpha particles?
- What do they do to matter?

- What are beta particles?

- What stops beta particles?

- What are gamma rays?

- How can radioactivity be detected?
- What is half-life?

- Is there a safe level of radioactive radiation?

- Give some safety precautions for handling radioactive materials?

... the nuclei of unstable (radioactive) atoms.

... fast moving positively-charged helium nuclei=2 protons+2 neutrons.

... a sheet of paper.

... because of their 'large' size and double + charge, they transfer their energy to matter in a short distance and cause intense ionization.

... high-speed negative (beta −) or positive (beta +) electrons.

... most can be stopped by a few millimetres of aluminium. Because of their smaller mass beta particles cause less ionization than alpha particles.

... short wavelength, high-frequency electromagnetic waves (photons). The photons have no charge and so cause less ionization than alpha or beta particles.

... with a G-M tube connected to a scaler or ratemeter.

... the time it takes half of a sample of radioactive atoms to decay.

... no – any amount of radiation can damage the body. The higher the level of radiation, the greater the risk of short-term or long-term disease.

... use long tongs, keep the materials in lead-lined boxes, store in remote corners, never point them at the eyes or sex organs, never take them into the mouth or breath the dust.

Examination questions

1 (a) A G-M tube depends on the ionizing effects of radiation for its operation. Explain what is meant by 'the ionizing effects of radiation'.
(b) Radium is a radioactive material. Small radium sources are used in school laboratories. How should such a radium source be stored?
Explain how this storage method reduces the danger to people working near it.

(MEG)

2 (a) The common isotope of carbon ($^{12}_{6}C$) has a mass number of 12 and an atomic number of 6.
Describe the structure of an atom of this isotope of carbon.
(b) How does an atom of $^{12}_{6}C$ differ from an atom of $^{14}_{6}C$?
(c) $^{14}_{6}C$ is radioactive. What does *radioactive* mean?
(d) Why are radioactive substances dangerous if absorbed by the human body?
(e) Describe briefly one use of a radioactive isotope in archaeology, or medicine. Make clear the principle involved in the use you describe.

(MEG)

3 In a factory which makes floor covering, changes in the thickness of the floor covering are detected using a radioactive source and a detector. The source emits beta particles.
(a) Describe how changes in the thickness of the floor covering are detected.
(b) Explain why neither alpha nor gamma sources would be suitable for this application.

(MEG)

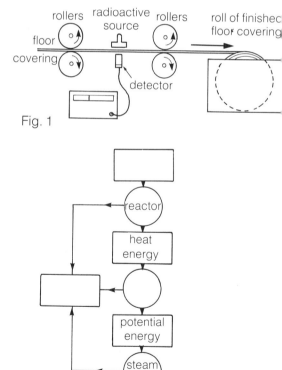

Fig. 1

4 (a) Underline the sources of energy in the list below which are renewable sources:
coal, waves, nuclear, gas, solar, geothermal.
(b) Fig. 2 shows energy conversions in a nuclear power station.
The *boxes* represent forms of *energy*.
The *circles* represent *devices* which convert energy into another form.
Copy and label the unfilled boxes and circles.
(c) State any two problems caused by a nuclear power station and for each problem suggest a suitable precaution.

(NEA)

Fig. 2

5 Fig. 3 represents the testing of the thickness of aluminium foil at a factory.

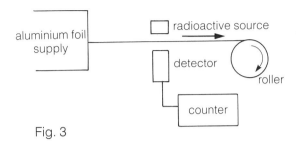

Fig. 3

The radioactive source gives out beta particles. It is placed above the sheet of foil and the detector below the sheet.
(a) From which part of the atoms of the source have the beta particles come?
(b) Why are alpha particles unsuitable for this test?
(c) The readings on the counter for a time of eight seconds are given in the table below.

Time in seconds	0	1	2	3	4	5	6	7	8
Total count	0	40	80	120	165	210	255	295	335
Count in one second	—	40	40	40	45	45			

(i) Copy and complete the table.
(ii) Examine the completed table. Now describe what is happening to the thickness of the foil, giving reasons for your answer.

(NEA)

6 An experiment was carried out to measure the density of steel ball bearings. This was done by dropping a known number of the ball bearings into water in a measuring cylinder and recording the increase in the water level. Fig. 4 shows the water level before 200 ball bearings were added and Fig. 5 shows the water level after they were added.
(a) Read the level in Fig. 4 (initial) and level in Fig. 5 (final) of the water and hence calculate the volume of 200 ball bearings.
(b) Before the ball bearings were added to the water their mass was calculated by measuring the mass of an empty beaker and the mass of the same beaker containing the ball bearings. The following readings were taken:

  Mass of empty beaker = 215 g
  Mass of beaker + 200 ball bearings = 599 g

Calculate the mass of the ball bearings.
(c) Use your values for the volume and the mass of the ball bearings to calculate the density of the steel.
(d) Why do you think a more accurate value for density can be obtained by using 200 bearings rather than 20?

(NISEC)

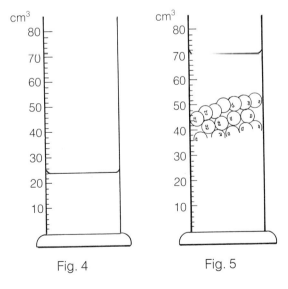

Fig. 4   Fig. 5

Answers

1 Forces
22 (a) (i) 60 m/s; (ii) 12 s

2 Measuring forces
2 (a) 350 N; (b) 24 cm
5 60 N

3 Forces in balance
13 P=491 N; Q=509 N

Examination questions
1 (iii) 1.3 N; (iv) 178 cm; (v) 0.09 N/mm
3 (b) 20 m/s
4 (a) (i) 12.5 N; (ii) 20 cm²; (iii) 0.625 N/cm²; (iv) 0.313 N/cm²
 (b) (i) 160 cm³; (ii) 7813 kg/m³
5 (a) 0.25 Nm; (b) 1.25 N; (c) 1.25 g

4 Measuring motion
5 (a) 2.8 s; 4.8 s; (b) 26 m
8 16.1 cm; 115 cm/s
10 20 m/s; 50 m; 12 s
12 3 cm/s; 25 cm/s; 100 cm/s
13 81 m
14 (b) 16.5 cm; (c) 1.65 cm/s
18 50 m/s, 233°; 13 m/s, 67°; 25 m/s, 323°; 18 m/s, 120°
20 (a) 5 m/s; (b) 7 m/s; (c) 6 m/s
23 15 m/s, 30 m/s²; 8 m/s, 4 m/s²; −20 m/s, −5 m/s²; 20 m/s, 20 m/s
24 40 m/s; 200 m
25 1.5 m/s; 9 m/s
26 8 s; 32 m
27 15 m/s; 30 s; 262.5 m

5 Force and motion
7 8 m/s²; 64 m/s
8 1.176 m/s²; 99.4 m
9 285 N; 3 m/s²
10 0.08 s
12 10 m/s²
14 8 m/s²
18 31.25 m, 0.77 s, 1.41 s; 25 m/s, 7.75 m/s, 14.1 m/s
19 0, 5, 20, 45, 80, 125 m
 0, 10, 20, 30, 40, 50 m/s

Examination questions
1 (a) 250 m; (b) 10 s; (c) 80 m; (d) 15 s; (e) 10 m/s; (f) 0
2 (b) (i) 2.5 m/s²; (ii) 20 m; (iii) 2.67 m/s
3 (a) (ii) 20 m; (iv) 15 m/s
 (b) (iii) 1400 kg
4 (a) 800 m/s²
 (b) (i) 0.067 s; (ii) 0.0267 m
5 (a) 5 km
 (b) (i) 10 m/s; (ii) 8 min 20 s
 (c) 2 m/s²
6 (a) 4 s; (d) 2400 N
7 (c) 25 cm

6 Pressure
3 100 kPa; 5 m²; 32 kN
4 3 N/cm²; 1200 N/cm²
6 30 kPa
7 10.1 m
8 41,800 kPa
9 99.7 N
31 45.5, 38.5, 33.3 kPa

Examination questions
1 (a) (i) 12.5 N; (ii) 20 cm²; (iii) 0.625 N/cm²; (iv) 0.313 N/cm²
 (b) 1200 cm³
 (c) 7813 kg/m³
2 (b) (ii) 328 K; (iii) 2 kPa

3 (a) (i) 5 N/cm²; (ii) 5 N/cm²; (iii) 5 N/cm²; (iv) 5 N/cm²; (v) 2500 N

8 Energy transfer
2 1000 J; 0.01 J; 3000 J; 1.25 J
3 40 J; 800 J
4 25 kJ
6 0.48, 0.32, 1.2 J
7 9000 J

9 Power
1 2 J; 0.5 W
 600 J; 60 W
 6000 J; 1000 W
2 200 W

10 Machinery
2 10, 1.2, 0.5 J
 9, 1, 0.45 J
 90%, 83.3%, 90%
9 2, 3, 4 m
 75%, 67%, 75%
10 30%, 25%, 5%, 80%

11 Wave energy
8 (a) 3; (b) $1\frac{1}{4}$; (c) 2; (d) up
9 (b) 7 cm
10 1 m/s, 3 m/s, 10 m/s, 11 m
 5×10^{14} Hz, 3×10^8 m/s, 300 m

12 Waves we can hear
6 250 Hz
9 200 Hz
16 715 m
28 100 m
29 2000 s

Examination questions
2 (a) 70 m
3 (c) (i) 1.5 cm; (ii) 15 cm/s
6 (b) (i) 4 m; (ii) 500 J; (iv) 840 J
9 (c) (i) 48 kJ; (ii) 24 s; (iii) 6 s
10 (b) (i) 100 s; (ii) 700 kJ; (iii) 7 kW

13 Light energy
10 2 m/s
13 9.25

16 The expansion of solids
9 4 min; 79%

Examination questions
2 (b) (ii) 3p
3 (e) 4.17 W

20 Electric energy
19 10^{12}
21 20 J, 1800 J, 400 s, 8192 J
35 2 waves

21 Cells and voltage
8 0.7 V, 3 V, 1.4 V
 400 mV, 70 mV, 3 V

22 Electric resistance
22 69°C
28 A=6 Ω; B=12 Ω; C=3 Ω; D=2 Ω; E=4 Ω; F=9 Ω; G=18 Ω
29 (i) 32 Ω; (ii) 6 Ω; (iii) 13.3 Ω; (iv) 2 Ω; (v) 3.2 Ω; (vi) 10.7 Ω
 (vii) 20 Ω; (viii) 8 Ω

32 A=3 A; B=9 A; C=4 A; D=8 A; E=6 A; F=2 A; G=4 A; H=6 A
33 (a) 6 Ω; (b) 5 A; (c) 2 A, 3 A
34 2 mA, 50 mA
    8.8 V, 4 V

23 Electric power
1 2198 J
2 20 W, 2 W, 10 W, 50 W
3 (a) 2 J, 10 J, 20 J
   (b) 50 s, 10 s, 5 s
5 40 W, 120 W, 2400 W, 960 W, 1 W
6 1/3 A, 3 A
   720 Ω, 240 Ω
   720 W, 240 W
8 3 kW, 1.5 kWh
   2000 W, 4 h
   0.05 kW, 0.5 h
   500 W, 0.5 kW
11 400 V, 0.05 A, 0.4 mA, 50 Ω
    625 kΩ, 80 kΩ, 0.6 W, 1.62 W
12 5 W; 0.42 A
13 (i) 2 A; (ii) 1.2 A, 0.8 A; (iii) 14.4 W, 9.6 W
14 1 W, 4 W
15 4 Ω

24 Electricity in the home
5 12 A, 4 A, 0.5 A, 3 A
   13 A, 13 A, 3 A, 13 A
   Examination questions
1 (a) 10 C; (b) 240 J; (c) 180 J
2 (a) (i) 1500 Ω; (ii) 2 mA;
   (b) 2 V
   (c) (i) 500 Ω
3 (b) (i) 1000 J; (ii) 4 A; (iii) 62.5 Ω
7 (b) (ii) 4.2 Ω, 9.8 Ω
   (c) 10 A, 10 W

28 Electricity generators
17 3 V, 15 V, 6 mV, 5 V
    1.2 s, 8 ms, 20 ms, 20 ms
    0.83 Hz, 125 Hz, 50 Hz, 50 Hz

29 Electric transformers
4 1250
10 (b) 200 W, 100 W
    (c) 100 W
15 350 V, 61 Hz
    Examination questions
2 (b) 4 V; (d) 2.5 ms; (e) 400 Hz
7 (a) (ii) 1600

32 Measuring matter
4 1 million
7 (a) 30; (b) 9 g
9 8 g/cm^3, 1 g/cm^3, 10 cm^3, 100 g
11 1.2 kg
12 60 m^3, 72 kg
13 8, 2, 0.5, 0.5

33 The change of state
18 380 J
24 702.4 kJ
26 1700 kJ
27 17 kJ
29 1000 s
30 240 kJ

35 Radioactivity and the atom
4 3600±60
5 21
11 3.8
14 1 min
   Examination questions
6 (a) 46 cm^3; (b) 384 g; (c) 8.4 g/cm^3

Index

absolute zero of temperature 65, 68, 158, 316, 318
absorption
 of light 127, 139
 of microwaves 100, 101
a.c. to d.c. 268
acceleration 11, 33, 50
 and force 40
 calculating 35, 41, 50
 negative 33, 50
 of gravity 41, 50
action 4
active sonar 108
air
 convection of 155
 cooling 167
 density of 65
 expansion of 155
 heating and cooling 65
 measuring expansion of 157
 pump 63
 resistance 11, 22
alpha rays 322, 324, 330, 337
alternating current 265
alternating voltage 191, 192, 279
 peak value of 271
alternators 265, 266
altimeter 62
ammeter 215, 242, 252
ampere 188, 194, 215
amplitude 111, 115
ancient fuels 78, 79
AND gate 290, 292
aneroid barometer 62, 68
anode 189, 191, 194
Antarctic 247
aperture control 136
Arctic 246, 247
atmospheric pressure 55, 60, 61, 62, 68
atoms 185, 320
atom bomb 328
attraction
 of charges 180
 of magnets 246
average speed 28, 50
axis 17

back projection 137
back voltage 287
background radiation 322, 329, 330
balancing 17, 18, 19
 a beam 18
 point 19
ballcock 155
barometer 62
base
 current 285, 293
 of a transistor 285
bathroom scales 14
battery 195, 196
 car 196
 charger 197
 hydrometer 197
 lead-acid 196
 top-up 198
bells 258

Bequerel 321
bequerel 332
Beryllium nucleus 320
beta rays 323, 324, 330, 337
bicycle reflector 131
bike alternator 267
bimetal strip 150, 173
black
 as a radiator 163, 173
 as an absorber 163, 173
 cooling fins 167
blow 48
bodies and radioactive radiation 329
Bohr model of atom 320, 336
boiling 304
 energy needed for 312
Bourdon gauge 62, 64, 68
Boyle's law 64, 68, 319
brakes 57
Brownian motion 315, 336
brushes 261
buzzer 258

cable ratings 238
caesium-137 333
camera 135
cancer cells 329, 330
capacitors 181, 194
 circuit symbol 181
 smoothing 270
car
 brakes 57
 battery 197, 198
 radiator 167
 relay in 257
carbon resistor 210
carburettor 146
cartridge fuse 237
catapulted motion 34
cavity insulation 165
ceiling rose 234
cells 195, 196
 lead-acid 196
Celsius temperature 170, 174
 conversion to Kelvin 66, 68
centre of gravity 19, 20, 23
chains
 energy 81, 114
chain reaction 328
chance happenings 325
change of state 303, 317, 336
charge 178, 179, 185
 electron theory of 185
 force between 180
 inducing 186
 storing 181
 types 180
Charles' law 158, 319
chemical energy 72
Chernobyl 332
chest expanders 16
circuit
 diagram 199
 electric 199, 226
 symbols 199
 two-switch 201
clinical thermometer 171
clouds 306

cm/s 28
collector
 current 285, 293
 of a transistor 285, 293
colours
 from light 127, 128
 of the spectrum 128, 139
combined logic gates 291
commutator 261
compass 247, 252
compressed air 64, 68
compression waves 106, 107
concave reflectors 125, 126, 139, 162
condensation 306
conduction
 by air 160
 by water 160
 of heat 159, 164
 testing 161
conductors, electrical 181, 194, 207
consumer unit 233
contact points 274
convection
 currents 153, 164
 of air 156, 173
conversion of energy 80
cooling
 by evaporation 305
 fins 167, 174
coulomb 188, 194
critical angle 129, 139
 measurement of 130
Curies 321
current control 209
 at a junction 224
 of people 188
cycle
 alternator 267
 pump 63, 64, 68

Darlington pair 287
deflection
 electrostatic 190
 of electron beams 190
demagnetization 255
density 299
 equation for 300, 336
 of air 301
 units of 300
deviation of light 127, 139
dew 306
dice 326
diffraction 114
 of microwaves 101
 of water waves 99
diffusion 316, 336
dimmer control 209
diode 211
 and back voltage 274, 287
 and Ohm's law 219
 and rectification 268, 280
 circuit symbol 211
 light-emitting 212
direct current 263
displacement 31, 96
distance travelled 28
 from speed/time graph 30, 50

distance/time graph 27, 28
distilled water 198
dose
 equivalent 331
 radiation 331
double glazing 165
double insulation 235
drag 11
dry cell 191
dynamo 263

Earth 5
 electric potential of 182
 gravitational field strength 44
 magnetic field of 247
earth
 pin 235, 236
 wire 233
earthing 183, 194, 236
effective dose equivalent 331
efficiency of machines 91, 94, 115
effort 90
electric charge
 negative 180, 185
 positive 180, 185
electric energy
 negative 182, 184
 paying for 230
 positive 182
electric shock 240
energy 72
 electrical 226, 227
 gravitational 72, 78, 84
 heat 173, 141
 kinetic 73
 light 119, 139
 potential 72
 sound 74

fire-proof suits 168
fission 328
flex 237, 243
 rating 238
fluorescent tube lights 187
focal length of a lens 132
focus
 of light rays 125, 132
 of microwaves 100
 of radiation 162
force
 of gravity 5, 23
 vectors 3
forces 2, 6, 23
 and acceleration 40
 and motion 38
 and work 82
 between charges 180, 194
 between magnetic poles 246
 in balance 18
 measuring size of 13
 motor 260
 of muscles 14
 turning effect of 17
force-meter 13, 14
fossil fuels 79, 114
four-stroke engine 146
free-body diagrams 3, 10, 21, 23

342

free-fall 44, 46, 47
freezing mixture 308
freon 307
frequency 97, 105
   audible range 105, 114
   measuring for a.c. 272
friction 7, 8, 23
   decreasing 8
   increasing 9
   of the air 11
   opposes motion 10
full-wave rectification 271
fuses 214, 233, 237, 238, 239
fuse tester 200

g 44, 45
gamma
   rays 322, 324, 330, 337
   waves 100, 103
gas laws
   Boyle's 64, 319
   Charles' 158, 319
   pressure 66, 319
gas pressure 59, 62
   kinetic theory of 318
   natural 79
gas turbine 144, 173
gates, electronic 288, 289, 290, 292
gears 92, 93
gear box 93
Geiger-Müller tube 321, 322, 327, 337
genetic effects 331
geothermal energy 72, 78, 84
gold-leaf electroscope 182
gram 298
gravitational energy 72, 78, 84
   equation of 85, 114
gravitational field strength 45, 50
gravity
   acceleration of 41
   force of 5
   work against 84
gray 331

half-life 326, 327, 337
half-wave rectification 271
hard magnetic materials 280
hearing and frequency 105, 114
heat
   and change of state 303
   and work 141
   calculation of 310
   conduction of 159
   energy 141, 173
   engines 94, 142
   from nuclear fission 328
   measurement of 309
   radiation of 162, 164, 173
heating and cooling air 65
heating
   electric 212
   equation 311
helium nucleus 320
hertz (Hz) 105
hidden heat 304
high-voltage generator
Hiroshima 329
home, electricity in 233
home-made projector 120
Hooke's law 15, 23
hot and cold water system 154
hot air 65
house radiator 167
hovercraft 32
hydraulic machine 56, 58, 68
   motor powered 58
hydroelectric 78

hydrogen 320
hydrophones 108

ice point 174, 170
images
   formation of 119
   mirror 121, 139
   real 120, 134, 139
   virtual 120, 121, 124, 134, 139
impulse 48
incidence, angle of 139, 123
increasing friction 9
induced charges 186, 194
induction
   coil 274
   electromagnetic 273, 280
   magnetic 250, 280
inertia 47, 50
infrared waves 100, 103
insulators
   electrical 181, 194, 207
   of heat 160, 165, 173
   testing 161
interference of waves 99
inverse proportion 64
iodine-131 330, 333
ionization 325, 337
iron
   electric 151, 173
   filings 249
   magnetic properties of 248, 250
iron wire
   resistance change with temperature 220
isotopes 320, 337

jet engines 144
joule 80, 83
joule/coulomb 194
joulemeter 227
junction
   division of current at 224

keeping warm 164
Kelvin temperature scale 66, 68
kilogram 298
kilowatt-hour 230, 242
kinetic energy 73
   equation of 85, 115
kinetic theory of matter 314, 316, 317, 336
km/h 28
knots 10

lamps 199
   filament and Ohm's law 219
   in parallel 205
   in series 204
   power of 228
latching circuit 290
latent heat 304, 308, 336
   measurement of 312
law of moments 18, 21, 23
left-hand rule 260, 280
lens camera 135
   light control of 135
lenses 132
   concave 133
   focal length of 133
   focus of 132
   images 119, 133, 139
   strength of 132
levers 90
   light 123
light
   energy 119, 139
   and LDR's 221
   sensing circuits 224

   images 119, 139
   levers 123
   pipes 132
   rays 123
      deviation of 127
      dispersion of 127, 128, 139
   spectrum of 127
   waves 100, 103
light-dependent resistor (LDR) 286, 287
   and Ohm's law 220, 243
   relay control 257
   logic control 290
light-emitting diode (LED) 212, 242
   and Ohm's law 217
   protective resistance 219
lighting circuit 233
lightning 185
lithium nucleus 320
load 90
loft insulation 165
logic microchips 288
longitudinal waves 106, 107, 115
loudness of sound waves 111, 115
luminous body 119

machines 115
   efficiency of 91, 115
   gears and pulleys 92, 93
   hydraulic 56, 68
   lever 90
   wheel and axle 91, 115
magnet 247
magnetic fields 249, 280
   of a solenoid 253, 280
   of a current 252, 280
magnetic
   field lines 250
   induction 280
   poles 246, 254, 281
magnetism 246
magnetizing 253
magnifying glass 120, 139
mains
   alternator 266, 279
   electricity 233, 280
   frequency 192, 279
   switch 233, 234
   voltage supply 279
manometers 59, 68
marathon fever 188, 204, 208, 214, 224
mass 6, 7, 8, 23, 29, 336
matrix board 284
measuring cylinder 298
melting 308
microchips 288
microphone 111
microwaves 100, 101, 103
   uses of 101
mirror images 121, 124, 139
model
   of a gas 314
   of the atom 320
moments 17, 23
   balanced 18
   clockwise and anticlockwise 18
   law of 18
momentum 48, 50, 144
moon 6
Morse code 201
motion 25, 26
   and force 38
   Brownian 315
   equations of 36, 50
   first law of 38

   second law of 1, 4, 48
   third law of 4, 23
motor
   electric 261
   force 260
multimeter 217

Nagasaki 329
nails 10
NAND 290, 292
National grid 279
negative gates 290
neon lamp 179, 183, 187
neutrons 320, 328, 330, 336
Newton
   equation of motion 42, 50
   first law of motion 38, 50
   second law of motion 41, 48, 50
   third law of motion 4, 23, 50, 144
newton 2, 6, 13
newton-meter 22
Nichrome 208
NOR 290, 292
normal 123, 139
North pole 246
NOT gate 288, 290, 292
nuclear
   decay 325, 326
   fission 328
   power 328
nucleon number 320, 336
nucleus 320, 325
   of atom 185

ohm 217
Ohm's law 218, 242
   testing for 219
oil 79
   tanker 183
operating theatres 184
optical fibres 132
OR gate 289, 292
orbit
   humans in 47
ORP12 220, 242, 257, 286, 290
oscilloscopes 191, 194, 267, 269, 270
   measurements with 271, 280

parallel
   adding resistors in 222, 242
   lamps in 205
   particle beam, negative 189
particles of matter 314
pascal 53, 68
passive sonar 108
periscope 131
perspiration 305, 336
petrol
   engine 145, 173, 274
   tanker 184
photocells 78
photosynthesis 77, 114
pitch of a note 111, 115
pivot 90
plotting compass 250
plug, 'mains' 237
plumb line 19, 20
points, car 275
poles of a magnet 246, 260
   force between 246
popping corn 325
potential difference 195, 226, 231, 242
potential energy 72
   electric 178
potentiometer 210

343

power 86, 114
　electric 227, 229, 242
　of a bike 86
　of a bunsen 87
　of electric motor 87
　of legs 89
　of pendulum 73
　problems 88
practical pots 210
pressure 53, 68
　measuring 53
　in liquids 54, 68
　of atmosphere 60, 61
　of gas 318, 319, 336
　of sound waves 111
　through liquids 55, 68
　units of 53, 68
Pressure law 66, 68, 319
principle focus of lens 133
prism 127, 128, 139
　internal reflecting 130, 139
　uses of 131, 139
projector 120, 137
proportional 15, 16
protons 320, 336
proton number 320, 336
pulleys 92, 93
pump cycle 63
pumped storage system 89
push and pull 2

quality factor 331
quality of sound 111, 115

radiation sickness 329
radiators
　car 167
　house 167
radio waves 100, 102, 103
radioactive
　decay 325
　energy 79
radioactivity 320, 321
　dose 330
　man's use of 329
radium 321
radon 327, 330
rainbow colours 128
reaction forces 4, 23
real image 120, 134, 139
rechargeable cells 196, 242
recharging a battery 197
recording motion 26, 27
rectification 268, 280
　full-wave 271
　half-wave 271
reducing friction 8
reed relays 256
reed switches
　normally closed 255
　normally open 255
reflection
　angle of 123, 139
　law of 123, 126
　of radiation 162
　of waves 99
　total internal 129, 130
refraction
　of light 126, 139
　of waves 99
refrigerator 307
relay 256, 257, 280
　switch-off voltage 274
　transistor control 287
resistance
　electric 207, 226
　equation of 217
　internal 226
　measuring 216
　of iron with temperature 220

　of LDR with light 221
　of thermistor with
　　temperature 220
　units of 217
resistor 208
　adding 222
　carbon 210
　circuit symbol of 208
　extra special 211
　variable 209
right-hand grab rule 253
right-hand rule 265, 281
ring main 235
ripple tank 98
rising main 155
rocket engine 144, 145, 173
rotor 261
Rutherford 325

safety
　circuit 202
　electrical 239
　level of radioactivity 337
　radioactive 331, 337
salt/ice mixture 308, 336
satellite 78
scalars 31
scaler 321, 327, 337
seeing 124
seivert 331
semiconductor diode 219
series
　adding resistors in 223, 242
　lamps in 204
shiny receiver 163
short 214
shutter speeds 135
signal generator 105, 111
skid pad 9
slide projector 137
smoke 315
　cell 315
smoothing capacitor 270
sockets, electrical 235
soft magnetic materials 280
solar
　heating 78
　panels 78
solenoid 253
　field of 253, 280
　uses of 253, 280
sonar 108
sound waves 106
　in liquids 108
　in solids 107
　in vacuum 109
　shape of 111
South pole 246
space station 47
spark 179, 183, 184, 274, 322
　plug 146
specific heat capacity 310, 336
　of copper 309
　of water 310
specific latent heat
　of fusion 312, 336
　of vaporization 313, 336
spectrum of light 128, 139
speed 28, 50
　and direction 31
　and time 30
　control for motor 262
　measuring 29
　of light 102, 119, 126, 162
　of sound 110
　of waves 97
　terminal 11
　/time graph 30, 50
speedometer 30
spring balance 14, 23

spring of air 64
stability 19
　neutral 19
standard atmospheric
　　pressure 55
starter motor 258
states of matter 303
　kinetic theory of 336
steam point 170, 174
steam turbine 143
steel, magnetic property of 247,
　　248
stopping radiation 323
strain energy 72
straw, sucking with 61
streamline 11
stretching materials 15
stroboscope 98
sun's energy supply 77
switches 200
　transistor as 285
　twilight 286
　two-way 203
symbols 199
　capacitor 181
　diode 211
　fuse 214
　resistor 208

television 102, 103
temperature 169
　low 308, 336
terminal speed 11
terminals, negative and positive
　　195, 215
thermistor 290
　and Ohm's law 219, 220, 242
thermocouple 173
thermometer 169, 174
　alcohol 170
　and thermistors 22
　bimetal 152
　clinical 171
　electronic 171
　lower fixed point of 170
　upper fixed point of 170
thermopile 163
thermostat 151
　bimetal 173
　electronic 224
thoron 327
three-pin plug 235
throttle 146
ticker-tape 26
　to measure speed 27, 50
ticker-timer 26, 28
tidal energy 79, 114
time-base circuit 192
timers 26
toppling 20, 23
total internal reflection 129
　by prisms 130
transformers 272, 275, 280
　circuit symbol 277
　power of 276
　turns ratio 277, 280
transistor 284, 292
　as a switch 284
　npn 284
　pnp 284
　test circuit 284
transmission of pressure 56
transverse waves 106, 115
travelling electrons 187
truth tables 201, 289
tungsten lamp filament 213
turbine 141
　gas 141, 173
　steam 143, 173, 266
turning effect 17, 18, 23

twilight switch 286
two-switch circuits 201

U values 165
ultraviolet waves 100, 103
unit, electrical 230
unstable 19, 325
uranium 321, 325, 328

vacuum 61
　sound through 109, 115
vacuum flask 164, 174
vacuum tube 189, 191
valve 298
valves, petrol engine 146, 147
Van der Graaff generator 183,
　　187
vapour 305, 306
variable resistor 209
vectors 3, 23, 31
　adding 32
　diagrams of 32
　sum of 33
velocity 31, 50
　adding 32, 33
vent pipe 155
vibrations 95, 104
　of tuning fork 104
virtual image 120, 121, 134,
　　139
voltage 191, 192, 195
　control by heat 225
　control by light 225
　control by temperature 225
　control of 225
　dividing up 225
　step-down 278
　step-up 274, 278
　supply 198
voltmeter 195, 198, 242
volts 182, 192, 194
volume 298

washer
　cycle pump 63
water cooling 167
watts 227
wave equation 97, 114
wavelength 95, 97, 111, 114
waves 95
　amplitude 115
　diffraction 1, 14, 99
　electromagnetic 100, 103,
　　144
　pictures of 96, 119
　reflection of 99
　refraction 99
wax, melting and freezing 308
weather 78
weight 6, 7, 23, 336
weightlessness 46, 50
wheel and axle 91, 92, 115
windmills 78
wire
　fuse 237
　live 233, 280
　neutral 233, 280
work 82, 114
　amount of 82
　against gravity 84, 114
　measurement of 83, 114

X-plates 191, 192
X-rays 100, 103
xenon 333

y-amplifier 192
y-plates 191

zero, electric potential 183